SELF-CONTROL, DECISION THEORY, AND RATIONALITY

Thinking about self-control takes us to the heart of practical decision making, human agency, motivation, and rational choice. Psychologists, philosophers, and decision theorists have all brought valuable insights and perspectives on how to model self-control, on different mechanisms for achieving and strengthening self-control, and on how self-control fits into the overall cognitive and affective economy. Yet these different literatures have remained relatively insulated from each other. *Self-Control, Decision Theory, and Rationality* brings them into dialog by focusing on the theme of rationality. It contains eleven newly written essays by a distinguished group of philosophers, psychologists, and decision theorists, together with a substantial introduction, collectively offering state-of-the-art perspectives on the rationality of self-control and the different mechanisms for achieving it.

JOSÉ LUIS BERMÚDEZ is Professor of Philosophy and Samuel Rhea Gammon Professor of Liberal Arts at Texas A&M University.

SELF-CONTROL, DECISION THEORY, AND RATIONALITY

New Essays

EDITED BY

JOSÉ LUIS BERMÚDEZ

Texas A&M University

CAMBRIDGE
UNIVERSITY PRESS

CAMBRIDGE
UNIVERSITY PRESS

University Printing House, Cambridge CB2 8BS, United Kingdom

One Liberty Plaza, 20th Floor, New York, NY 10006, USA

477 Williamstown Road, Port Melbourne, VIC 3207, Australia

314–321, 3rd Floor, Plot 3, Splendor Forum, Jasola District Centre, New Delhi – 110025, India

79 Anson Road, #06–04/06, Singapore 079906

Cambridge University Press is part of the University of Cambridge.

It furthers the University's mission by disseminating knowledge in the pursuit of education, learning, and research at the highest international levels of excellence.

www.cambridge.org
Information on this title: www.cambridge.org/9781108420099
DOI: 10.1017/9781108329170

© José Luis Bermúdez 2018

First published 2018

Printed and bound in Great Britain by Clays Ltd, Elcograf S.p.A.

A catalogue record for this publication is available from the British Library.

Library of Congress Cataloging-in-Publication Data
NAMES: Bermudez, Jose Luis, editor.
TITLE: Self-control, decision theory, and rationality : new essays / edited by Jose Luis Bermudez, Texas A & M University.
DESCRIPTION: Cambridge, United Kingdom ; New York, NY : Cambridge University Press, 2018.
IDENTIFIERS: LCCN 2018027623 | ISBN 9781108420099
SUBJECTS: LCSH: Self-control – Philosophy. | Decision making – Philosophy.
CLASSIFICATION: LCC BF632 .S524 2018 | DDC 153.8–dc23
LC record available at https://lccn.loc.gov/2018027623

ISBN 978-1-108-42009-9 Hardback

Contents

Figures

Tables

Contributors

ARIF AHMED is University Reader in Philosophy at the University of Cambridge.

CHRISOULA ANDREOU is Professor of Philosophy at the University of Utah.

JOSÉ LUIS BERMÚDEZ is Professor of Philosophy and Samuel Rhea Gammon Professor of Liberal Arts at Texas A&M University.

KENNY EASWARAN is Associate Professor of Philosophy at Texas A&M University.

NATALIE GOLD is Senior Research Fellow in the Faculty of Philosophy at the University of Oxford and Principal Behavioral Insights Advisor at Public Health England.

LEONARD GREEN is Professor of Psychological and Brain Sciences and of Economics at Washington University in St Louis.

ALFRED R. MELE is William H. and Lucyle T. Werkmeister Professor of Philosophy at Florida State University.

JOEL MYERSON is Research Professor of Psychological and Brain Sciences at Washington University in St Louis.

MARTIN PETERSON is Professor of Philosophy and Sue and Harry Bovay Professor of Engineering Ethics at Texas A&M University.

HOWARD RACHLIN is Distinguished Professor of Psychology at Stony Brook University.

REUBEN STERN is Postdoctoral Fellow at the Center for Mathematics and Philosophy at the Ludwig-Maximilian University of Munich.

JOHANNA THOMA is Assistant Professor of Philosophy, Logic, and Scientific Method at the London School of Economics.

PETER VALLENTYNE is Florence G. Kline Chair in Philosophy at the University of Missouri.

PAUL WEIRICH is Curators' Distinguished Professor of Philosophy at the University of Missouri.

Acknowledgments

The papers in this collection were all presented at the Self-Control, Decision Theory, and Rationality Workshop held at Texas A&M University on May 15–16, 2017. The workshop was funded by a grant from the Philosophy and Psychology of Self-Control Project at Florida State University, itself funded by the John Templeton Foundation. I am grateful to the Philosophy and Psychology of Self-Control Project and the Templeton Foundation for their financial support, to the contributors for the timeliness and quality of their contributions, and to the Department of Philosophy at Texas A&M University for financial and administrative help. Finally, it has been a pleasure to work with Hilary Gaskin and the team at Cambridge University Press.

Introduction

José Luis Bermúdez*

Self-control raises fundamental issues at the heart of practical decision making, human agency, motivation, and rational choice. Unsurprisingly, therefore, it has been studied and discussed within several different academic literatures. Psychologists, philosophers, and decision theorists have all brought valuable perspectives on and insights into how to model self-control, different mechanisms for achieving and strengthening self-control, and how self-control fits into the overall cognitive and affective economy. Yet these different literatures have remained relatively insulated from each other.

The chapters in this collection bring those literatures and approaches into dialog by focusing on the rationality of self-control. This Introduction begins by comparing and contrasting the different approaches to self-control taken in philosophy, psychology, and decision theory, respectively. After setting up a schematic and illustrative puzzle of self-control as a framework for mapping out the different contributions, I then draw out some principal themes running through the different chapters and briefly introduce each contribution.

Philosophers, psychologists, and decision theorists typically approach the topic of self-control in subtly different ways. These approaches are in many ways complementary, but it will be helpful to begin by sketching out some characteristics of the different disciplinary perspectives on self-control before drilling down more deeply into the themes and arguments of the chapters in this collection. (*Warning:* I will be painting selectively with broad strokes of the brush.)

I.1 Philosophy: The Greek Background

Philosophical engagement with self-control dates back at least as far as Socrates, and subsequent discussion (at least within the Western tradition)

---

* Work on this Introduction was supported by a grant from the Philosophy and Psychology of Self-Control Project funded by the John Templeton Foundation.

has been very much circumscribed by ways of thinking about self-control directly traceable to Socrates, Plato, and Aristotle.[1] Ancient Greek discussions of self-control were framed by the guiding idea that self-control is a virtue, or at least a character trait, typically defined by contrast with its opposite, weakness of will.

The Greek word *akrasia* that has now become standard terminology for talking about weakness of will only appeared in mainstream philosophical discussion in the work of Aristotle, but the phenomenon itself was discussed extensively by Socrates and Plato. In the dialog *Protagoras*, Socrates counterintuitively and provocatively denied that weakness of will, as standardly construed, exists. The qualification is important. Socrates was not denying the existence of weak-willed behavior. He was not denying that people often have a second glass of wine or a third slice of cake when they think it best that they abstain. His objection was to standard ways of thinking about weakness of will. In particular, he denied that weak-willed behavior comes about when one's knowledge of what one ought to do is somehow overruled by the pleasures of the moment. Knowledge, for Socrates, cannot be "a slave, pushed around by all the other affections."[2] And so, by the same token, self-control cannot be a matter of resisting temptation and, more generally, mastering the emotions.

In the *Republic*, Plato gave an exceptionally clear articulation to precisely the conception of self-control and weakness of will that Socrates had rejected. For Plato, the space for self-control is set by ongoing conflict between the rational and irrational parts of the soul (*psuche*). The irrational parts of the soul dominate in the weak-willed person, while self-control results when the rational part of the soul prevails. Aristotle's condensed and often difficult to understand discussion in the *Nicomachean Ethics* incorporates elements from both the Socratic and the Platonic perspectives but definitely comes out closer to Plato than to Socrates.

In one respect, though, all three of the great Greek philosophers are in agreement. They each take weak-willed behavior to be a paradigm of practical irrationality and self-control to be a rational ideal. For Plato and Aristotle, the practical rationality of self-control is achieved through the

---

[1] The essays in Bobonich and Destrée (2007) extend the discussion of *akrasia* through the later Greek philosophers up to Plotinus. Outside the Western tradition, self-control has been discussed by classical Indian and Chinese philosophers. In the Nyâya dualist tradition, self-control (*svâtantrya*) was taken by some philosophers to be a distinguishing mark of the conscious (see Chakrabarti 1999: chap. 7), while the control of the emotions was a recurrent theme in early Chinese philosophy, as discussed in Virág (2017).

[2] Plato, *Protagoras*, 352 B.C., trans. W. K. C. Guthrie. In Hamilton and Cairns (1961: 344).

exercise of willpower – a conscious effort to resist temptation and master affects and emotions. For Socrates, in contrast, weak-willed behavior is the result of an agent miscalculating the overall pleasures and pains that will result from a given course of action – and so, correlatively, the practical rationality of self-control comes from correctly applying what he calls the *art of measurement* (correctly assessing the balance of pleasures and pains in every action and situation).[3]

This focus on practical rationality is the hallmark of most subsequent philosophical discussions of self-control. Such a focus is typically aligned with a model of practical reasoning and agency that invokes intentions and judgments about what is all-things-considered better. The focus tends to be on explaining how all-things-considered judgments can either prevail or be overruled. How do we need to think about the human mind and the human motivation system to understand how self-control can take place and how it can fail?[4]

This way of thinking about self-control is often synchronic. That is, the discussion is of how, at a given moment of choice, it is possible for the passions to be mastered, temptation overcome, and better judgment vindicated. But some philosophers, particularly those influenced by discussions in decision theory (on which see further later), have explored how self-control can play out in a diachronic context, where the problem is how an agent can adhere to earlier resolutions and commitments when motivations change, both when such motivations are anticipated and when they are not.[5]

I.2 Psychology: Mechanisms of Weakness and Mechanisms of Control

From the perspective of psychology, the focus has tended to be on mechanisms – both the mechanisms that make self-control necessary (the mechanisms of weakness, as it were) and the mechanisms that make self-control possible (the mechanisms of strength). Considerations of rationality have not typically been at the forefront of discussion, but the emphasis on

---

[3] Or at least that's how he describes things in *Protagoras*, where he derives this view from a version of psychological hedonism. Scholars disagree about whether the psychological hedonism in *Protagoras* is Socrates' considered view or simply a dialectical tool that suited his purposes at the time. The former interpretation is defended in Irwin (1995: §60, pp. 85–87).

[4] For recent, influential, and representative discussions, see Holton (2009) and Mele (2012).

[5] See, for example, the essays by Michael Bratman collected in Bratman (1999).

finding techniques for enhancing self-control often suggests an implicit assumption that self-control is practically irrational.

Pioneering studies of animals and humans have shown that the need for self-control can be conceptualized in terms of different ways of discounting the future. Very few subjects, human or animal, value goods to the same degree irrespective of the time that they will be received. Typically, the value a good is perceived to have decreases more the longer the time before it is received. The rate at which value diminishes with temporal distance is a function of how the subject discounts the future. This is particularly relevant to a class of cases that will be discussed further later, namely where good intentions are thwarted by temptation. What happens when agents backtrack on commitments (breaking a diet, for example, or not following through on an exercise program or a savings plan)? When the plan to diet, to exercise, or to save is made, the perceived long-term benefits of following it outweigh the anticipated, but still distant, fleeting rewards of back-sliding. And yet, when those fleeting rewards (the extra cake, the lie-in, or the extra disposable income) are near at hand, temptation can overcome the best resolutions. How should this kind of preference reversal be understood?

Influential studies by George Ainslie, Howard Rachlin, and others have shown that psychological phenomena such as these can be understood in terms of discount functions that have a particular form. These are discount functions where the rate of change varies over time. In what are known as *exponential discount functions*, the discount rate remains constant. This means that a given delay, say a day, will be accorded the same weight whenever it occurs. So, for example, if I prefer $10 today to $11 tomorrow and I discount the future in an exponential manner, then I will prefer $10 in 100 days to $11 in 101 days. Most people do not take this approach, however. Even if I am not prepared to wait until tomorrow for an extra dollar, the same reward would probably lead me to extend my wait from 100 days to 101 days. This pattern of preferences can be understood in terms of discount functions that are *hyperbolic*. In a hyperbolic discount function, the rate of discounting is affected by the (temporal) proximity of the item being discounted. As we will see in more detail later, this means that agents with hyperbolic discount functions can succumb to preference reversals. Psychologists have explored the relation between discounting and various types of addictive and compulsive behavior.

Preference reversals are not inevitable. Temptation does not always win out. But what are the mechanisms that make this possible? Outside the laboratory, people often talk about self-control in terms of the exercise of willpower, often treated as a kind of psychic force that some people have more than others and that can be cultivated and strengthened. This basic idea goes back at least as far as Freud, but one influential development within scientific psychology of this very intuitive idea is the ego depletion theory, originally due to Roy Baumeister and Dianne Tice.[6] According to ego depletion theory, willpower is a limited resource than can be used up and run down completely. In the influential experiment that launched the theory, students who had held back from freshly baked chocolate chip cookies and instead snacked on radishes gave up much sooner on a tricky geometric puzzle than either a control group or students who had eaten the cookies. As with a number of areas of social psychology, however, doubts have been raised about the experimental support for ego depletion theory.[7]

Ego depletion theory makes much of the image of self-control as being like a muscle both in so far as it can be used to exhaustion and in so far as it can be strengthened through use. This second idea is independent of the first, and a number of experiments seem to show that regular "exercise" can strengthen willpower. So, for example, Megan Oaten and Ken Cheng found that subjects who pursued a two-month physical exercise regime did better than control subjects on standard laboratory self-control tasks.[8] Other possibilities explored for improving the mechanisms of self-control include implementation intentions (developing determinate strategies in advance for dealing with specific temptations) and defusing temptation by representing the object of temptation in a "cool" rather than "hot" motivationally charged way (or, alternatively, representing the long-term gain in a hot rather than cool way).[9]

I.3 Decision Theory: Problems of Dynamic Choice

Classical decision theory codifies the conception of instrumental rationality dominant in the social sciences, most prominently in economics (excluding subfields of economics such as behavioral economics and experimental economics, which explicitly explore alternatives to standard models of rationality). Decision theorists typically model instrumental

---

[6] See Baumeister et al. (1998), and for a more popular presentation, see Baumeister and Tierney (2011).
[7] A multilaboratory replication project sponsored by the Association for Psychological Science found little evidence for the ego-depletion effect (Hagger et al. 2016).
[8] Oaten and Cheng (2006). Relatedly, see Muraven et al. (1999) and Muraven (2010).
[9] For a review of research into implementation intentions, see Gollwitzer and Sheeran (2006). Walter Mischel originated the hot/cool systems approach. See, for example, Mischel and Ayduk (2004) and Mischel et al. (2011).

rationality (for individual decision makers in nonstrategic situations) through some form of expected utility theory. Popular versions of expected utility theory all make the same basic prescription, which is that a rational decision maker will choose an option that maximizes utility when the utility assigned to an option's outcomes is appropriately weighted by the probability that it will occur. In other words, rational decision makers maximize expected utility.

From the perspective of classical decision theory, self-control presents something of a puzzle. This is because exercises of self-control are typically acts (or omissions – when self-control leads me not to act) that refer back to an antecedent decision or commitment. So, for example, my staying up late preparing my class is an exercise of self-control by virtue of my commitment to being adequately prepared for my class. But for this to count as an act of self-control, it would seem that the agent's prior decision or commitment cannot motivationally outweigh the agent's current desires and preferences. If my desire to be the designated driver is stronger than my desire for a drink, then how am I exercising self-control in declining the drink? Surely, I am just doing what I most want to do. So, in instances of self-control, the agent's current desires and preferences will typically outweigh her prior decision or commitment.

This is problematic for classical decision-theoretic accounts of dynamic choice (sequences of choices) because a rational decision maker will only take into account her utility assignments at the moment of choice, ignoring any earlier assignments not reflected in her current assignments. This is often called the *historical separability of preferences* (or the *time separability of preferences*).[10] So, if the agent has undergone a preference reversal (so that the fleeting temptation has become more attractive than the long-term goal), the separability of preferences seems to make completely irrelevant the high utility previously assigned to reaching the long-term goal. On one interpretation of the time separability of preferences, classical decision theory prescribes that rational decision makers should choose *myopically*, looking only at the here and now and ignoring how they have valued things in the past.

Against that prescription some decision theorists have proposed that rational agents in such a situation need either to be *sophisticated choosers* or *resolute choosers*.[11] A sophisticated chooser in effect "ties herself to the mast," like Odysseus preparing himself to sail near the Sirens. A sophisticated chooser

[10] See McClennen (1990) for a comprehensive and influential discussion of separability assumptions in decision theory.
[11] Sophisticated choice strategies originate with Strotz (1956) but the terminology with Hammond (1976).

might give her car keys to a friend, for example, or agree to forfeit money if she misses a workout. Such precommitment strategies are intended effectively to remove the option of succumbing to temptation. The sophisticated chooser does not exercise self-control at the moment of choice. She exercises it in advance, reasoning backward in a way that eliminates what she believes (anticipating her future preferences) to be nonfeasible options. Sophisticated choice remains consistent with the time separability of preferences because the sophisticated chooser looks forward but reasons backward.

A resolute chooser, in contrast, eschews the precommitment strategies of sophisticated choosers.[12] He sticks to his guns at the moment of choice, conforming to his earlier plan even in the face of temporarily reversed preferences. The resolute chooser has nonseparable preferences because he is swayed by previous valuations that are motivationally outweighed at the moment of choice. Because the time separability of preferences is a natural extension of expected utility theory when it is applied in a dynamical context, this means that the resolute chooser will not be an expected utility maximizer.[13]

The concept of resolute choice is the closest that decision theory comes explicitly to modeling self-control. And yet it raises two fundamental questions. The first is how (if at all) the rationality of resolute choice can be defended within the instrumental perspective of decision theory. There is at least a prima facie tension between instrumental models of rationality and the everyday phenomenon of self-control. The second is how resolute choice is even possible. The discussion in the decision-theory literature has focused primarily on the rationality of resolute choice. The actual mechanisms have received little to no discussion in that literature.

These two questions relate closely to the issues respectively explored in the preceding two sections. Clearly, ongoing discussions in philosophy, psychology, and decision theory explore intersecting and overlapping topics. The chapters in this volume break ground in drawing those discussions together. I turn now to setting up a schematic puzzle of self-control that will allow me to introduce some of the principal themes of the individual contributions.

I.4   A Paradigm Case of Self-Control

Drawing on some of the threads emerging in the preceding section, I will define a paradigm case where self-control seems to be required. This case

---

[12]   For a defense of resolute choice, see McClennen (1990, 1998). See also Gauthier (1997).

[13]   See McClennen (1990: §7.5, esp. n. 7).

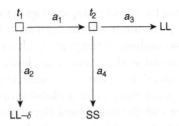

Figure I.1 The paradigm case of self-control represented as a sequential choice problem. The moment of planning is at time t_1 with the moment of choice at time t_2. At t_1 the agent has a choice between making a precommitment to LL (which would guarantee receiving LL, but at a cost, namely δ) or continuing to t_2. At t_2 the choice is between SS and LL.

raises some fundamental questions about the nature, exercise, and rationality of self-control. I will then go on to situate the chapters in this volume through the different answers that they offer to these questions.

The paradigm case that I am proposing is similar to standard cases discussed in the psychological and decision-theoretic literatures. Let us assume that an agent makes at time t_1 a commitment or resolution to pursue a large, long-term benefit at a later time t_3. At a time t_2, later than t_1 and earlier than t_3, the agent has the opportunity of abandoning the long-term commitment in favor of a small, short-term reward. I follow standard practice of using the abbreviations LL (for *larger, later*) and SS (for *smaller, sooner*). At the time of making the resolution, the (discounted) value of LL is more powerfully motivating than the (discounted) value of SS. That is to say, the utility that the agent assigns at time t_1 to the future receipt of LL is greater than the utility she assigns to the future receipt of SS. However, by time t_2 the agent's preferences have (temporarily) reversed, and now SS motivationally outweighs LL. Because t_2 is the moment of choice, this is an opportunity for the agent either to exercise self-control or to succumb to temptation and weakness of will.

We can depict this paradigm case as a sequential choice problem, as illustrated in Figure I.1. In addition to the option at time t_2 of exercising self-control and holding out for LL instead of succumbing to SS, Figure I.1 represents the sophisticated choice option of adopting some sort of pre-commitment strategy. The outcome of the sophisticated choice option (a_2) is the long-term reward LL minus the cost of precommitment, represented by δ. So, to complete the dynamic choice typology sketched out here, option a_4 represents myopic choice (weakness of will or succumbing to temptation), while option a_3 is the self-controlled, resolute choice.

This schematic decision problem raises three distinguishable, but definitely interrelated, sets of questions. The first set of questions clusters around the concept of rationality. Most obviously, one might ask how a rational agent should tackle this decision problem. Most of the contributors to this volume are working within (or at least exploring the consequences of) a broadly instrumental decision-theoretic conception of rationality, so they typically approach the issue by thinking about how, if at all, instrumental theories of rational choice tackle decision problems, such as our paradigm case, that seem to require the exercise of self-control. The chapters by Thoma (Chapter 1), Peterson and Vallentyne (Chapter 2), and Weirich (Chapter 3) are all situated within this general area, exploring the claims of decision theory to provide a standard of rationality for decision problems of the basic type of the paradigm case just sketched out (often extending the discussion to more complicated decision problems).

A second set of questions clusters around the mechanisms responsible for the basic preference reversal that gives the decision problem its force. How does a space for temptation arise, even in the face of a strong resolution? What can we learn about self-control and how to exercise it from studying what makes weakness of will possible? And, moreover, issues of rationality arise here too. Is it really the case, as is often assumed, that the psychological phenomena that create a space for weakness of will betoken a degree of practical irrationality? Questions such as these come to the fore in the chapters by Ahmed (Chapter 4), Green and Myerson (Chapter 5), Rachlin (Chapter 6), and Andreou (Chapter 7).

Finally, a third set of questions clusters around the mechanisms that potentially lead to self-control – to taking option a_3 and holding out for LL rather than choosing myopically and opting for a_4 and SS. Even if you think that a rational agent will, all other things being equal, hold fast to commitments in the face of temptation (as opposed to either giving into temptation or adopting a sophisticated precommitment strategy), the question still arises as to how that can actually happen, given the agent's motivational profile at time t_2, the moment of choice. The chapters by Bermúdez (Chapter 8), Mele (Chapter 9), Gold (Chapter 10), and Easwaran and Stern (Chapter 11) present a range of different perspectives on this important question.

The remainder of this Introduction uses this general mapping of the theoretical landscape to introduce the individual contributions.

I.5   Rationality, Dynamic Choice, and Self-Control

Can it be rational to exercise self-control to resist temptation? As suggested earlier, this is a puzzling question. On the one hand, agents seem to do better in the long run and by their own lights if they do exercise self-control – which suggests that self-control is instrumentally rational. On the other hand, though, self-control requires overriding one's current desires and for that very reason seems to be instrumentally irrational. Chapters 1 through 3 present different perspectives on this puzzle.

In "Temptation and Preference-Based Instrumental Rationality" (Chapter 1), Johanna Thoma evaluates two of the most frequently canvassed lines of argument in this area. The first line of argument employs a two-tiered strategy, with exercises of self-control counting as instrumentally rational when the level of evaluation is shifted from individual actions to deliberative strategies. David Gauthier's analysis of a sequential prisoner's dilemma modeled on Hume's famous example of the two farmers at harvest time is a good example of how this might work.[14] But, according to Thoma, all such strategies are doomed to fail because the shifted preferences that are problematic at the level of individual actions simply reappear at the higher level of deliberative strategies (because strategies that permit selective exceptions to accommodate temptation will typically be preferred to strategies that do not).

The second line of argument is exemplified by Ned McClennen's discussion of resolute choice (see footnote 12). What makes resolute (i.e., self-controlled) choice instrumentally rational, he argues, is that the resulting plan is Pareto superior to its sophisticated and myopic alternatives when evaluated over the agent's successive selves (time slices) – in other words, it improves things for some time slices without leaving any time slices worse off. The problem with this approach, Thoma maintains, is that it imposes irreconcilable demands on successive time slices. On the one hand, they must be sufficiently unified to care about each other's preferences and resolutions, but on the other, they must be sufficiently independent of each other that the preferences of the current time slice do not immediately trump those of the other time slices. Thoma sees no way of combining these demands.

Both lines of argument share a common assumption about rationality, namely that the standard for assessing how instrumentally rational a given

---

[14] See Hume (1739/1978: III.2.5, 520–21) for the original example and Gauthier (1994) for the two-tiered strategy.

choice is must be set by the agent's preferences. Thoma terms this the assumption of *preference-based rationality*. Perhaps abandoning the assumption will solve the problem? Abandoning the assumption (which Thoma thinks that there are anyway good reasons to do) means holding that there is a true standard of instrumental rationality that may be misrepresented by an agent's preferences. But it now becomes impossible to give an argument that resisting temptation is instrumentally required because whether or not it is rational to resist temptation will be a function of whether the true standard of instrumental rationality is stable or not, which is something that instrumental rationality itself cannot prescribe (because it can reason only about means, not about ends).

Martin Peterson and Peter Vallentyne's "Self-Prediction and Self-Control" (Chapter 2) approaches the rationality of self-control from a somewhat different angle. Instead of asking whether and how it can be rational to exercise self-control, Peterson and Vallentyne tackle the question of how we should model the choices of agents who have rational preferences and beliefs but only limited self-control over their future actions (and who are aware that their self-control is limited). The authors operate with a distinction between rational preferences and motivational preferences (thereby illustrating one way of moving beyond what Thoma calls the assumption of preference-based rationality) so that an agent's motivational preferences (the springs of action) may not be aligned with his rational preferences (which reflect his values and deliberation). How should you behave if you know (or have good reason to believe) that you may be susceptible to this sort of misalignment?

Well, resolute choice is not an option, according to Peterson and Vallentyne. Resolute choice requires (and does not explain) perfect self-control, so it cannot be sensitive to, or provide guidance for, anticipated failures of self-control. Instead, they hold, a rational agent with limited self-control is best off adopting a version of the sophisticated choice strategy through some form of precommitment or the like. But Peterson and Vallentyne have a new take on how to understand the reasoning that leads to sophisticated choices. It is generally assumed that sophisticated choosers must employ some form of backward induction (i.e., starting with the last decision node of a sequential choice problem and working backward to the current node), but, as they observe, backward-induction reasoning cannot be applied in all cases. It fails, for example, when there is no identifiable last node. So they propose an alternative – a version of sophisticated choice on which agents can rationally assign probabilities to their future choices and then maximize expected utility relative to those probability assignments.[15]

Peterson and Vallentyne respond to a number of objections to this conception of sophisticated choice. Isaac Levi, Wolfgang Spohn, and others have argued that there is no stable way of reasoning probabilistically about one's future actions (because, as Levi puts it, deliberation crowds out prediction).[16] In response, Peterson and Vallentyne describe various ways in which one can ascribe probabilities to future actions without thereby influencing those actions. They also consider two objections that apply to standard (backward-induction) versions of sophisticated choice. The first is that sophisticated choosers breach a form of dominance principle, and the second is that sophisticated choosers are committed to being dynamically inconsistent.

Paul Weirich's "Rational Plans" (Chapter 3) in effect poses the question of whether the possibility of rational self-control can ever come into conflict with standard theories of rational choice. In other words, does the (rational) exercise of self-control require a decision maker to do anything other than maximize (expected) utility? On the face of it, the rational exercise of self-control involves choosing a sequence that maximizes even though individual acts within that sequence do not maximize. Weirich denies that this is a correct description of the situation, however. A person who rationally exercises self-control will always (and ipso facto) be maximizing either utility (in cases of decision making under certainty), expected utility (in decision making under risk), or expected utility as computed using some admissible pair of utility and probability assignments (in decision making under uncertainty).

Weirich starts by considering ideal decision makers operating under certainty in a decision problem where every node has an option with maximum utility. The decision makers know where they are in the decision tree. They have full information about their history. And they can predict their choices at future nodes. Such decision makers will be sophisticated choosers, and at each decision node they will maximize utility and thereby produce the best sequence through the tree (as Weirich illustrates with a backward-induction argument). There is no opportunity, therefore, for an agent to choose rationally while failing to maximize. This conclusion sets the baseline for subsequent discussion, and Weirich proceeds by

---

[15] As they observe, there are affinities between sophisticated choice, as they understand it, and wise choice, as proposed by Rabinowicz (e.g., Rabinowicz 1995).
[16] See Levi (1989) and Spohn (1977).

showing that the same conclusion holds even when we move from the baseline case to consider decision problems where (1) some sequences of choices are not directly compared, (2) some branches are infinitely long, and (3) the decision tree includes (a) chance nodes or (b) nodes for other agents.

In considering cases (1) through (3b), Weirich emphasizes that the rationality of acts has priority over the rationality of sequences. Only things that are under the agent's direct control can be assessed for rationality using the standard of utility maximization, and sequences of choices are not under the agent's direct control. For this reason, the rationality of sequences of choices derives from the rationality of the individual choices that go to make them up. Weirich considers (and rejects) three candidates for consistency constraints on sequences of choices additional to the standard of utility maximization. These include putative requirements of dynamic consistency, conditionalization of preferences, and coherence. None of these requirements can, in Weirich's view, trump the requirement that each choice in a rational sequence maximizes utility. And, as he discusses with reference to Arntzenius, Elga, and Hawthorne's example of Satan's apple, cases of rational self-control can lead to acts that are irrational (because there are alternatives that have greater utility).[17] But because those acts are irrational, they do not pose a threat to the general principle that rationality requires utility maximization. For Weirich, therefore, an agent can rationally exercise self-control even though the resulting act is itself irrational.

I.6 Preference Reversal, Time Inconsistency, and Impulsivity

The opposite of self-control goes by a range of different names. It is variously called *weakness of will, impulsivity, incontinence, akrasia,* or *succumbing to temptation,* with different terminology often highlighting different aspects of a complex phenomenon. Four of the chapters in this collection concentrate on understanding this complex phenomenon, thereby deriving lessons for how we think about self-control.

Arif Ahmed's "Self-Control and Hyperbolic Discounting" (Chapter 4) starts off by identifying three basic features of weakness of will, which he

---

[17] Satan divides an apple into infinitely many pieces and offers them one by one to Eve, who derives utility from each one but will be banished from the Garden of Eden if she takes them all. It is rational for her to bind herself in advance to stop after a certain number of pieces, say 617. This would be rational self-control. But when she gets to the 617th piece, stopping at that point is irrational (because she would maximize by taking it). See Arntzenius, Elga, and Hawthorne (2004).

models in terms of competing desires. First, preferences are reversed over time. Second, the desire that wins out is the more impulsive one. And third, the two desires are genuinely in conflict. In order to capture these basic features, he treats weakness of will as a phenomenon of future discounting (very much in line with how the phenomenon is often viewed within psychology, as discussed earlier). He distinguishes two basic types of discount functions. Some discount functions are delay consistent. That is, a given interval of time is discounted to the same degree irrespective of how near it is to or far it is from the present. Discount functions that lack this property are delay inconsistent. Whereas delay-consistent discount functions are always exponential, delay-inconsistent functions are typically hyperbolic. Some time-inconsistent discount functions are *impulsive*, in that they treat a delay as less important the further in the future it occurs. Ahmed equates weakness of will with impulsivity of the discount function, which both accommodates and goes some way toward explaining the three basic features of weakness of will.

Ahmed sees an intriguing parallel between cases of temptation and the cluster of puzzles collectively termed *Newcomb's problem*. Newcomb's original problem, together with various alleged real-life instances, is widely thought to illustrate the shortcomings of a purely evidential approach to decision theory because these are cases where actions are causally irrelevant to the outcomes that are correlated probabilistically with them (e.g., where my desire to smoke is caused by a lesion that also causes lung cancer so that desisting from smoking will have no effect at all on my health). If we assume, as may seem plausible, that no individual act of resisting temptation will causally affect my strength of will, then I seem to be in a Newcomb-type situation when I am confronted with temptation. Resisting temptation will be good evidence that I am a strong-willed person but will not actually bring it about. So evidential versions of decision theory will prescribe self-control, while causal versions will endorse impulsivity. Ahmed, in line with his previous work on Newcomb cases and evidential decision theory, takes this to be an argument in favor of evidential decision theory.[18]

From a public-policy perspective, Ahmed observes that impulsive discounting often involves foreseen behavior that decision makers know they will regret. For this reason, an impulsive discounter can be rationally motivated to bind his future self (to be a sophisticated chooser) – for example, when the current discounted value of the temptation is negative.

---

[18] For Ahmed's more general discussion of evidential decision theory, see Ahmed (2014).

This opens up the possibility of liberal rather than paternalistic policy approaches to protecting people who are tempted by gambling and other impulsive behaviors. Government bodies can set up precommitment mechanisms that would enable gamblers, for example, to limit themselves in advance to an upper bound on either the time they spend gambling or the losses they incur (on the model of the YourPlay scheme introduced by the government of the state of Victoria in Australia). This would count as liberal rather than paternalistic because it would be optional. Moreover, the model of incontinence as impulsive discounting makes it possible to calculate the optimal time for making available a precommitment option and also to fix the amount that the decision maker should be prepared to pay in order to precommit.

Leonard Green and Joel Myerson's "Preference Reversals, Delay Discounting, Rational Choice, and the Brain" (Chapter 5) takes a somewhat different approach to the complex relations between self-control, weakness of will, discounting, and impulsivity. They propose a modification to the standard model of discounting, suggesting that a hyperboloid function fits the data better than a hyperbolic function, particularly when it comes to discounting at long delays.[19] But, even with this modification, they caution against identifying impulsivity with a discount function of a particular shape. The relation between preference reversal and hyperboloid discounting is also more complex than usually assumed. So, for example, there are amount effects in discounting, with smaller delayed rewards discounted more steeply than larger ones. And, in fact, when the size of the reward is factored in, it is possible for exponential discount functions to predict preference reversals.

According to Green and Myerson, moreover, the data speak against the idea that steep discounting reflects some personality trait of impulsivity or lack of self-control. Substance abusers are widely viewed as epitomizing lack of self-control, and there is, as they observe, considerable empirical evidence both that substance abusers discount the future much more steeply than control subjects and that they discount their substance of choice more steeply than, say, money. But still, there is at best a weak correlation between this type of steep discounting and standard measures of impulsivity in experimental tasks and personality tests. Moreover, there is an asymmetry in how substance abusers discount. They discount delayed

---

[19] Let V be the current value of a delayed reward A, with D the length of the delay. Then an exponential discount function typically takes the form $V = Ae^{-kD}$, where k is a rate parameter and e is Euler's number. A hyperbolic discount function might be presented as $V = A/(1 + kD)$ and a hyperboloid as $V = A/(1 + kD)^s$, where s is an additional scaling constant.

rewards more steeply than control subjects but show no difference from control subjects when it comes to delayed losses or probabilistic rewards.

With respect to the mechanisms of self-control, Green and Myerson take issue with the idea that steep discounting necessarily reflects an irrational lack of *willpower*. Quite apart from the empirical objections to willpower models of self-control (as described earlier in the section on psychological approaches to self-control), it may well be that steep dis-counting makes good sense (as it does for a small bird in winter, whose high metabolic rate makes it most unwise to hold out for the chance of a larger, later reward). They also explore the role of conscious experience in self-control, particularly the experiential imagining of different outcomes. Green and Myerson cite data from patients with amnesia whose disorder prevents them from imagining the future. The data show that such patients show no difference in delay or probability discounting from control sub-jects, which they take to suggest that decision making about the future does not require the ability to (re-)experience or imagine any specific event in the past or future

Howard Rachlin's "In What Sense Are Addicts Irrational?" (Chapter 6) explores the relation between discounting and self-control and impulsivity from the perspective of a general contrast between neurocognitive and behavioral-economic models of rationality, themselves derived from asso-ciated models of action. Behavioral-economic models focus on contingen-cies between behavior and consequences, and the basic currency of behavioral-economic models, as it were, is long-term patterns of behavior rather than consistency in choices. From the perspective of the behavioral-economic model, types of discount functions (e.g., hyperbolic versus exponential) or the steepness of discount curves does not reveal irration-ality either in addicts or in normal subjects. Rather, rationality is a matter of creating patterns in your behavior to maximize value in the long run. Addicts are irrational to the extent that they fail to create such patterns.

Rachlin explores the analogy between selfishness and impulsiveness and the general question of whether there is a relation between rationality in self-control and rationality in social cooperation. He describes some of his own experiments on social discounting, in which participants were asked to contemplate sharing hypothetical amounts of money with people at varying degrees of social distance. It turns out that a hyperbolic discount-ing function provides a good fit to the data. Moreover, there is a significant correlation between delay discounting and social discounting. Rachlin extrapolates from this to a more general hypothesis about the nature of addiction.

Addicts treat their future selves as they would treat people far from them in social distance ... The irrationality of addicts may be most usefully understood to lie, not in the description of their choices by one sort of discount function or another, but in their failure to identify with their past and future selves. That is, even when she is nominally playing a game against her future self, the addict plays it as a non-addict would play a game against other people, socially distant from her.

From the perspective of Rachlin's behavioral-economic approach to rationality, what matters are long-term patterns. So he suggests that self-control is improved by strategies and mechanisms that focus attention on the long-term consequences of particular choices. In particular, he emphasizes what he terms *soft commitment*, which is really a way of patterning choices, illustrating it with pigeon studies. When pigeons are allowed to choose between pecking two buttons, one giving a single food pellet immediately (SS) and the other giving four pellets with a four-second delay (LL), they end up pecking the SS button nearly 100 percent of the time. But when each button needs to be pressed twenty times, the same pigeon will peck the LL button (even though after nineteen pecks it is in the same position as in the original trial). This response persistence is what he calls *soft commitment*. The human equivalent (which Rachlin tested in a much more complex monetary experiment with delayed payoffs) is persisting in a choice (following a rule) that leads to a larger long-term reward once that choice has originally been made, focusing attention on the long-term pattern rather than on the individual choices of which the pattern is made up. This has implications for treating addiction, suggesting the strategy of teaching addicts behavioral methods for patterning choices.

In "Why Temptation?" (Chapter 7), Chrisoula Andreou opens up a novel way of thinking about clashes between temptation and self-control. As with the other authors described in this section, she focuses on the mechanisms that can lead an agent routinely to succumb to temptation. But, unlike them, discounting the future is not her primary concern. Accepting that cases where agents routinely succumb to temptation in the face of their own long-term goals are failures of instrumental rationality, she considers the question of how such failures can occur in a motivational system that is well designed for goal-directed agency. The view she defends is that susceptibility to temptation is not in fact an anomaly (due to lack of willpower or distortions in how the future is discounted) but rather an inevitable side effect of a motivational system generally well suited to the task of balancing multiple and competing goals. Her proposal has the virtue of covering cases in which there are no preference reversals, the temptations are not in any obvious sense a source of pleasure, and the agent's ultimate goals are not susceptible to criticism.

The characteristic feature of a motivational system susceptible to temptation is that it permits a mismatch between an agent's assessment of the importance of a goal and the goal's motivating force. This mismatch allows temptations to override long-term goals to the contrary because the long-term goals' motivating forces are temporarily weakened. At the same time, though, it can emerge from a design that allows agents to balance competing goals. Andreou begins by considering a simple turn-taking motivational system (one that simply allows the agents to satisfy competing desires in sequence) and then supplements it with degrees of sensitivity to affordances that might present themselves as the agent navigates through the environment (e.g., the increased availability of a particular food source, even though it is "out of turn"). Affordance sensitivity creates a space for primes to operate and hence for agents to act counter to their long-term, stable motivations.

If agents were time-slice agents, then this would be a very bad thing. But, for extended agents, a motivational system that combines turn taking and affordance sensitivity can be a useful mechanism for allowing an agent efficiently to balance competing goals that cannot be simultaneously satisfied. Moreover, just as it creates a space for temptation, it also creates a space for exercising self-control. Primes can be created as well as found. Implementation intentions and other self-control strategies can function as primes. Andreou points out that even during stretches of time where goals and assessment of relative importance remain relatively stable (i.e., there are no significant ranking reversals), the operations of such a motivational system can generate behavior that intuitively seems to qualify as reflecting lack of self-control.

I.7 Mechanisms for Self-Control

How does self-control actually work? What mechanisms allow agents to resist temptation and hold out for long-term goals? Many of the chapters already discussed propose answers to these questions, but they are the main focus of the last four chapters in this volume.

In "Frames, Rationality, and Self-Control" (Chapter 8), I tackle the unmodified version of the paradigm case. As I formulate the problem, to exercise self-control in the paradigm case is to make what McClennen called a "resolute choice." McClennen, despite making a cogent case for the rationality of resolute choice, had little to say about the mechanisms by which it might be achieved. Richard Holton has proposed an account of strength of will in terms of particular types of intentions that he calls "resolutions." Resolutions are intentions formed for the specific purpose of resisting inclinations to the contrary (and in particular inclinations to reconsider the intention itself). Holton (2009: 141) suggests that his account of resolutions "might be thought of as providing philosophical underpinnings for McClennen's work." I object to Holton's account on the grounds, first, that it overintellectualizes the phenomenon of resisting temptation and, second, that it does not explain when it is (and is not) rational to reconsider an intention.

I propose an account of how self-control can be exercised in the paradigm case that, parting company with McClennen and Holton, adheres to the time separability of preference (i.e., without having to appeal to prior intentions or any desires or preferences except those active at the moment of choice). This paper is part of a larger project that takes issue with an almost universally held view that it is irrational for one's evaluation of an option or an outcome to be affected by how it is framed. In effect, my target is the view that practical reasoning is necessarily extensional. This view is almost universally held but rarely argued for explicitly. I accept that many of the experiments in psychology and behavioral economics suggesting that susceptibility to framing effects is both widespread and inescapable do indeed reveal irrationality. It is hard to make a case that it can be rational (to take the famous Asian disease experiment) to value the outcome where 200 of 600 people die differently from the outcome where 400 of 600 are saved. But, I observe, practical reasoning often presents choices far more complex than those in the experimental framing literature, and I offer general reasons for thinking that practical reasoning can be what I term an *ultraintensional context*. An ultraintensional context is one in which it can be (*note: can be*, not *must be*) rational to assign different values/utilities to what is presented as two separate outcomes, even when they are known to be identical and to differ only in how they are framed.

As applied to self-control, my idea is that one way of exercising self-control in the paradigm case is through reframing the LL outcome. So, for example, instead of just thinking about the long-term goal outcome simply as an end in itself (as, say, *delayed LL*), the agent can frame it partly in terms of how it is achieved (as, say, *having successfully resisted SS*). If, as I propose, practical reasoning is an ultraintensional context, then that opens the door for the agent to prefer *having successfully resisted SS* to *taking SS*, even though the agent also prefers *taking SS* to *delayed LL*. Such an agent would certainly have quasi-cyclical preferences (in virtue of knowing full well that *having successfully resisted SS* and *delayed LL* are different ways of framing the same outcome). How could such quasi-cyclic preferences possibly be rational? I argue that preferences are rational when they are based on reasons, and I offer three different types of reason that are brought into play when the LL outcome is framed as *having successfully resisted SS*. Exercising self-control through framing the LL outcome can work by allowing agents to engage with, and be motivated by, types of reason that might otherwise be invisible to them.

Alfred R. Mele's "Exercising Self-Control: An Apparent Problem Resolved" (Chapter 9) takes issue with a prima facie puzzle about the possibility and rationality of a paradigm case of self-control that I proposed to the contributors to this volume when inviting them. The paradigm case is basically the sequential choice problem discussed above in section I.4. Mele modifies the case that I originally presented. His version of the paradigm case is as follows:

> The paradigm case occurs when at a time t_1 an agent judges it best on the whole (from the perspective of his own values and beliefs) to pursue a large long-term benefit (LL) at a later time t_3 and makes at t_1 a commitment or resolution to pursue LL at t_3. At a time t_2, later than t_1 and earlier than t_3, the agent has the opportunity for a small short-term reward (SS). Although at the time of making the judgment and resolution his desire to pursue LL is motivationally stronger than his desire to pursue SS, and although at t_2 the agent believes that it is better to pursue LL than SS, by t_2 SS motivationally outweighs LL. Getting either of LL or SS precludes getting the other.

Mele's principal modification is to introduce the notion of the agent's all-things-considered judgment about what it is best to do. This is very much in line with standard philosophical discussions of self-control (which, of course, Mele himself has done much to shape). It's worth remarking, however, that Mele's reformulation arguably removes the paradigm case from its original context, which had to do with how self-control might be conceptualized within the context of decision theory. Many decision theorists would hold that there is no room in the theory for a notion of all-things-considered best judgment over and above the agent's utility function.

In any event, Mele makes his reformulated version of the paradigm case more challenging by accepting the following principle:

> *Principle T* Whenever we act intentionally, we do, or try to do, what we are most strongly motivated to do at the time.

Prima facie, Principle T makes it impossible to exercise self-control in the paradigm case. After all, the case is set up so that at time t_2 the agent is most strongly motivated to pursue SS, which would seem to make the all-things-considered best judgment in favor of LL powerless. Mele rejects this conclusion, however. He emphasizes (as is standardly assumed in discussions of action theory) that it is perfectly possible for an agent to perform more than one intentional action at a time. This is important because (although he does not put it in quite these terms) exercising self-control is a temporally extended process. An agent can intentionally follow the dictates of temptation while at the same time setting in motion a process that will ultimately result in his resisting temptation.

Here is Mele's example of how this might work: Ian is sitting on the couch watching TV. His all-things-considered best judgment is that he should stop watching after the next commercial break and then go back to working on the house (LL). But, as things stand, he is most strongly motivated to stay put with the TV on (SS). Principle T tells us that Ian will stay on the couch watching TV. But, as Mele points out, this is perfectly compatible with Ian simultaneously embarking on a process that will eventually lead him off the couch – by, for example, uttering to himself a self-command to go back to work. In uttering the self-command (or adopting some other comparable strategy), Ian is inspired by his all-things-considered better judgment, even though (in accordance with Principle T) he follows his strongest motivation at the time and remains (for the moment) firmly in place on the couch. What makes this possible, Mele holds, is the basic difference between the motivational strength of our desires, on the one hand, and how we evaluate the objects of our desires, on the other.

The final two chapters in this volume approach the mechanisms of self-control in terms of cooperation between different temporal stages (time slices) of a single agent. In "Putting Willpower into Decision Theory: The Person As a Team Over Time and Intrapersonal Team Reasoning" (Chapter 10), Natalie Gold applies to self-control her work on team reasoning in game theory, developed in collaboration with Michael Bacharach and Robert Sugden. Many decision theorists think of sequential choice problems such as the paradigm case in terms of interactions between transient selves-at-a-time – an assumption that is, of course, closely tied to the time separability of preferences. And Gold thinks that this approach is independently plausible. As she puts it, "[t]he time-slice model captures the idea that choices are made at a time, by selves who have a first-person perspective on that time and a third-person perspective on the past and the

future." Moreover, it opens up the possibility of applying to the intrapersonal problem of self-control models developed to solve problems of
interpersonal collaboration, such as team reasoning.

The motivation for team reasoning comes from well-documented
difficulties in game theory. The principal solution concept in classical
game theory is Nash equilibrium, and there are many games in which
the intuitively rational and collaborative solution is not a unique Nash
equilibrium. In games such as Stag Hunt and Hi Lo, there are multiple
Nash equilibria with no principled way of choosing between them,
whereas in games such as prisoner's dilemma, the collaborative solution
is not a Nash equilibrium at all. Proponents of team reasoning think
that the problem lies with the highly individualistic conception of
rationality adopted by the vast majority of game theorists. Instead,
they propose a different model, one that shifts both the level of agency
and the way that agents think about payoffs. In team reasoning, agents
move from thinking of themselves as individual agents to thinking of
themselves as components of a team agent. Correspondingly, they shift
from thinking in terms of individual payoffs to thinking in terms of
group payoffs.

How can team reasoning be applied to self-control? Gold favors
Bacharach's model of what he terms "circumspect team reasoning,"
which takes into account the probability that other agents (time slices)
will engage in team reasoning.[20] Gold shows that team reasoning typically
leads to self-control in temptation cases that are construed as threshold
public good cases, which are cases in which a good is made available to
a group only if cooperation exceeds a certain threshold. So, in the example
Gold considers, an agent has three days to prepare for an examination, with
her three time slices each facing a choice between studying and not studying and each preferring not studying to studying but passing the examination to failing. The threshold for passing the examination is studying for
two days. In this case, circumspect team reasoning will lead the agent's first
time slice to study when she believes that the probability of the second time
slice studying is greater than 3/7. As Gold observes, successfully bringing
team reasoning to bear on self-control is conditional on an account of what
leads individuals to team identify with other time slices in a way that
creates the person-over-time as a team agent. This might emerge from joint

---

[20] Bacharach's theory is developed most fully in his posthumously published book, *Beyond Individual Choice: Teams and Frames in Game Theory* (Bacharach 2006), edited by Natalie Gold and Robert Sugden.

recognition that they all share long-term interests – or from a sense of similarity between proximate selves.

Kenny Easwaran and Reuben Stern's "The Many Ways to Achieve Diachronic Unity" (Chapter 11) also tackles the issue of how different time slices can cooperate over time, but they approach it from the opposite perspective. They focus on elucidating ways in which cooperation can emerge in the absence of what many philosophers have held to be its necessary psychological and normative prerequisites. Michael Bratman, for example, has argued that agents' unity over time (their capacity for diachronic consistency, for being self-governing over time, and for exercising self-control) is due to their being able to form future-directed intentions that are binding.[21] The efficacy of these future-directed intentions in binding agents to secure consistency is a function, says Bratman, of diachronic norms of consistency that rule out revising plans without appropriate reason. In particular, Easwaran and Stern discuss the norm that they and Bratman term D. According to D, it is locally irrational to intend at time t_1 to do something at a later time t_2, to take one's grounds to continue supporting doing that thing during the interval between t_1 and t_2, and yet to abandon one's intention at time t_2. Easwaran and Stern take issue with D on two grounds. First, they claim that it can only explain a relatively small proper subset of the cases where agents achieve self-control and diachronic consistency. Second, they offer examples where they think that adhering to D yields irrational consequences.

Easwaran and Stern propose a different approach to thinking about consistency over time, starting from the obvious fact that there are costs as well as benefits to reopening deliberation and reconsidering intentions. Often the question of whether or not it is rational to be consistent over time is fixed not by putative global norms such as D but rather by considerations of expected utility, where the relevant expectation is fixed, at least in part, by the agent's prior choices. In this vein, they consider three different schematic models in which a form of diachronic unity can be achieved without bringing into play norms of diachronic coherence or binding intentions. All these schematic models feature highly disunified selves – essentially collections of time slices, each of which has its own highly localized interests.

The first type of model that Easwaran and Stern consider treats the time slices of an individual as themselves individual agents, each of which has to choose whether to spend the next minute engaged in activity A or activity

---

[21] See, e.g., the essays in Bratman (1999).

B. Each agent has a utility function that is sensitive to what previous agents have chosen and that has a pair of structural properties. First, every agent has concave-up preferences. That is, there is an increasing marginal return to engaging in either of the two activities – each additional minute on activity *A* produces more value than any previous minute, and likewise for activity *B*. Each agent also prefers one activity to the other strongly enough that even spending only half the time doing that activity is preferred to completing the other (although, of course, they would prefer if all time slices engaged in their preferred activity). Easwaran and Stern show that such a disunified subject can nonetheless act in a way that gives the appearance of unity. For each utility function with this pattern, there will always be some number *i*, they show, such that if the first *i* agents have chosen *A*, then a later agent with that utility function will assign a higher expected utility to continuing with activity *A*, even if he happens to prefer the unanimous choice of *B* to the unanimous choice of *A*. Moreover, every such utility function will assign a higher utility to allowing a single agent to make a binding choice for all agents (irrespective of what that choice is) than they assign to choosing for themselves in the light of previous choices – provided that the dictatorial option is made available before they know how any of the other agents have chosen. The chapter builds on this basic (and, the authors admit, somewhat psychologically unrealistic) model by incorporating the costs of switching and the costs of introspection. In all of these cases, they show, what might seem from the outside to be the result of self-governance, self-control, and binding intentions can emerge without the psychological machinery that Bratman and many others have held to be essential for achieving diachronic unity.

This volume presents a variety of complementary and interlocking perspectives on how to think about self-control within different models of practical rationality, particularly those informed by decision theory. I hope that bringing them together will inspire further interdisciplinary work from philosophers, psychologists, and decision theorists on a topic of great theoretical and practical interest.

References

Ahmed, A. 2014. *Evidence, Decision, and Causality*. Cambridge: Cambridge University Press.
Arntzenius, F., A. Elga, and J. Hawthorne. 2004. Bayesianism, infinite decisions, and binding. *Mind* 113:251–83.

Bacharach, M. 2006. *Beyond Individual Choice: Teams and Frames in Game Theory*. Princeton, NJ: Princeton University Press.

Baumeister, R. F., E. Bratslavsky, M. Muraven, and D. M. Tice. 1998. Ego depletion: Is the active self a limited resource? *Journal of Personal and Social Psychology* 74:1252–65.

Baumeister, R. F., and J. Tierney. 2011. *Willpower: Rediscovering the Greatest Human Strength*. New York, NY: Penguin.

Bobonich, C., and P. Destrée. 2007. *Akrasia in Greek Philosophy: From Socrates to Plotinus*. Leiden: Brill.

Bratman, M. 1999. *Faces of Intention*. Cambridge: Cambridge University Press.

Chakrabarti, K. K. 1999. *Classical Indian Philosophy of Mind: The Nyâya Dualist Tradition*. Albany, NY: Southern University of New York Press.

Gauthier, D. 1994. Assure and threaten. *Ethics* 104:690–721.

Gauthier, D. 1997. Resolute choice and rational deliberation: A critique and a defense. *Nous* 31:1–25.

Gollwitzer, P. M., and P. Sheeran. 2006. Implementation intentions and goal achievement: A meta-analysis of effects and processes. In *Advances in Experimental Social Psychology* (pp. 69–119). New York, NY: Academic Press.

Hagger, M. S., N. D. L. Chatzisarantis, H. Alberts, et al. 2016. A multilab preregistered replication of the ego-depletion effect. *Perspectives on Psychological Science* 11(4):546–73.

Hamilton, E., and H. Cairns. 1961. *Plato: The Collected Dialogues*: Princeton, NJ: Princeton University Press.

Hammond, P. J. 1976. Changing tastes and coherent dynamic choice. *Review of Economic Studies* 43:159–73.

Holton, R. 2009. *Willing, Wanting, Waiting*. New York, NY: Oxford University Press.

Hume, D. 1739–40/1978. *A Treatise of Human Nature*. Oxford: Oxford University Press.

Irwin, T. 1995. *Plato's Ethics*. Oxford: Oxford University Press.

Levi, I. 1989. Rationality, prediction, and autonomous choice. *Canadian Journal of Philosophy* 19:339–63

McClennen, E. F. 1990. *Rationality and Dynamic Choice*. Cambridge: Cambridge University Press.

McClennen, E. F. 1998. Rationality and rules. In *Modeling Rationality, Morality, and Evolution*, ed. P. A. Danielson (pp. 13–40). New York, NY: Oxford University Press.

Mele, A. R. 2012. *Backsliding: Understanding Weakness of Will*. New York, NY: Oxford University Press.

Mischel, W., and O. Ayduk. 2004. Willpower in a cognitive-affective processing system. In *Handbook of Self-Regulation: Research, Theory, and Applications*, ed. R. F. Baumeister and K. D. Vohs (pp. 99–129). New York, NY: Guilford Press.

Mischel, W., O. Ayduk, M. G. Berman, et al. 2011. "Willpower" over the life span: decomposing self-regulation. *Social Cognitive and Affective Neuroscience* 6: 252–56.

Muraven, M. 2010. Building self-control strength: Practicing self-control leads to improved self-control performance. *Journal of Experimental Social Psychology* 46: 465–68.

Muraven, M., R. F. Baumeister, and D. M. Tice. 1999. Longitudinal improvement of self-regulation through practice: Building self-control strength through repeated exercise. *Journal of Social Psychology* 139:446–57.

Oaten, M. and K. Cheng. 2006. Longitudinal gains in self-regulation from regular physical exercise. *British Journal of Health Psychology* 11:717–33.

Rabinowicz, W. 1995. To have one's cake and eat it: Sequential choice and expected utility violation. *Journal of Philosophy* 92:586–620.

Spohn, W. 1977. Where Luce and Krantz do really generalize Savage's decision model. *Erkenntnis* 11:113–34.

Strotz, R. H. 1956. Myopia and inconsistency in dynamic utility maximization. *Review of Economic Studies* 23:165–80.

Virág, C. 2017. *The Emotions in Early Chinese Philosophy*. Oxford: Oxford University Press.

CHAPTER 1

Temptation and Preference-Based Instrumental Rationality

Johanna Thoma*

1.1 Introduction

We all seem to be prone to temporary shifts in our preferences. I might plan, for instance, to only stream one episode of a TV show during my coffee break. But then, once I have watched the first, I come to prefer to watch another. In the dynamic choice literature, such cases have come to be known as "temptation problems" because I can be said to be tempted to watch a second episode of TV.[1] Temptation problems confront us with a puzzle about instrumental rationality. On the one hand, an agent seems to do better by own lights if she does not give into the temptation and does so without engaging in costly commitment strategies. This seems to indicate that it is instrumentally irrational for her to give into temptation. On the other hand, resisting temptation also requires her to act contrary to the preferences she has at the time of temptation. But this seems to be instrumentally irrational as well.

My starting point here is that to make any progress in the resolution of this puzzle of instrumental rationality, we need to be more explicit about what we take to be the standard of instrumental rationality against which an agent's actions are evaluated. This chapter argues that it is a pervasive

* I thank all participants of the Rationality and Self-Control Workshop at Texas A&M University, as well as the Templeton Foundation for generously funding the workshop and José Bermúdez for organizing it and editing this volume. Previous drafts of this chapter also benefited greatly from detailed feedback from Sergio Tenenbaum, Michael Bratman, Jonathan Weisberg, Joseph Heath, Julia Nefsky, Peter Vallentyne, and an anonymous referee, as well as from discussion with audiences at the University of Toronto, the London School of Economics, the University of Bristol, the University of Kent, the University of Oxford, and the University of Copenhagen. I am grateful to the Balzan Foundation for funding my research stay at Stanford University, where the ideas in this chapter were first developed.
[1] See, for instance, Gauthier (1996) and McClennen (1998). Bratman (1998) and Holton (2009) consider the same kinds of problems but speak primarily of "evaluative judgments" rather than preferences.

but usually implicit assumption in rational choice theory that the agent's preferences over the objects of choice form the standard of instrumental rationality against which the agent's actions are evaluated. I call this assumption *preference-based instrumental rationality*. With this notion of instrumental rationality in hand, I consider the two most prominent types of arguments for why resisting temptation could be instrumentally rational, even though it requires us to act counterpreferentially. I argue that both arguments fail under preference-based instrumental rationality.

The first type of argument is a two-tier argument whereby not the agent's individual actions but her deliberative strategies over time are assessed instrumentally. Individual actions, in turn, are judged instrumentally rational if they are endorsed by the best deliberative strategy. A strategy that has the agent resist temptation is then argued to be instrumentally best. The core problem for two-tier accounts is that preference-based instrumental rationality implies that in temptation cases the standard of instrumental rationality itself shifts. I argue that this means that we can no longer say that a strategy of resisting temptation is instrumentally best. According to the second type of argument, resisting temptation is the result of mutually beneficial cooperation between the agent's *time slices*. Agents then have the same kinds of reasons to engage in this intrapersonal cooperation as they have to engage in interpersonal cooperation. I argue that given preference-based instrumental rationality, no plausible account of mutually beneficial cooperation between an agent's time slices can be given.

One might think that giving up preference-based instrumental rationality will help these arguments. However, I argue that this is not so. Doing so either doesn't do away with the problems, or it makes the arguments redundant, save for a special case. Giving up preference-based instrumental rationality creates the possibility that the agent's preferences misrepresent the true standard of instrumental rationality. But if the true standard of instrumental rationality is still shifting in a particular temptation case, the same problems as before arise. If, by contrast, the true standard of instrumental rationality is stable in a temptation case, then we already have a straightforward justification for why resisting temptation is instrumentally rational: it is best according to the true, stable standard of instrumental rationality.

The choice we thus face is the following: either we stick with preference-based instrumental rationality, in which case we are left to conclude that resisting temptation is instrumentally irrational – unless we find a better argument to the contrary – or we abandon preference-based instrumental rationality, in which case temptation cases may turn out to be much less puzzling. This chapter concludes by suggesting that the latter option makes both better sense of the phenomenon of temptation and has independent appeal.

1.2 A Temptation Case

Suppose that I like to stream an episode of a TV show when I take my afternoon coffee break. As my break starts, at t_1, I prefer to watch only one episode and then get back to work. But after I have watched that first episode, at time t_2, I prefer to watch another one over stopping. Once I have watched the second episode, I then return to my earlier preferences and would prefer just having watched the one episode.

Let O_0 be the outcome of watching no TV: I will get all my work done, but my coffee break will be boring. Let O_1 be the outcome of watching one episode, namely that I have an interesting coffee break and also get all my work done afterwards. O_2 is the outcome of watching two episodes: while I get to watch two episodes of an interesting show, I will not get my work done. Let $\succ$ represent strict preferences between outcomes. My preferences at the different points in time are the following:

$$t_1: O_1 \succ O_0 \succ O_2$$
$$t_2: O_2 \succ O_1 \succ O_0$$
$$t_3: O_1 \succ O_0 \succ O_2$$

The dynamic decision problem I face can now be illustrated by the decision tree in Figure 1.1. The square nodes here represent choices I need to make. In each case, I can decide whether to go "Up" or "Down."

Now suppose that if I make a decision at time t_2, I simply choose according to my preference between the available outcomes O_1 and O_2 and watch a second episode. Further suppose that I predict that I will do so at t_1 and treat this as certain. I then take myself to effectively face the choice between O_0 and O_2 at t_1. If, again, I simply go with my preference over these outcomes, I will choose to not watch any TV. I do so even though, at every point in time, I prefer watching one episode to watching no episodes.

As I said, this kind of case is often referred to as a *temptation problem* in the dynamic choice literature. The choice behavior I have just described, in turn, is referred to as *sophisticated*.[2] At each point in time, the agent predicts

---

[2] See McClennen (1990) for a formal treatment.

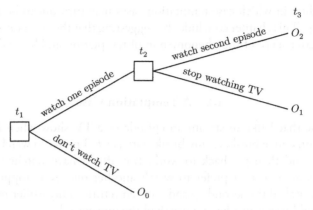

Figure 1.1 Temptation problem

what she will rationally do in the future. And, under conditions of certainty, she chooses in accordance with what act brings about the outcome she most prefers at that point in time, out of the ones that are still available to her then. If the agent treats her prediction of future rational behavior as certain, sophisticated choice simply follows from a rule that demands maximizing with regard to one's preferences at every point in time. Such a rule seems to be an archetypical requirement of instrumental rationality.

If I follow this sophisticated choice strategy in the temptation case, however, I will end up not watching TV. It is rationally impossible for me to watch one episode and then not give into the temptation to watch another. Yet many philosophers have thought that it must be rationally possible for agents to resist temptations of this sort. And then there must be something wrong with sophistication as a requirement of instrumental rationality or else resisting temptation is in conflict with instrumental rationality.

Indeed, one main source of the intuition that resisting temptation can be rational is itself instrumental in nature. And that is that the agent in these examples seems to end up worse off by her own lights. It seems like her life would go better if she had the capacity, at t_2, to not act in accordance with her temporary preferences. In fact, we can interpret the choice to not even watch the first episode as a kind of costly precommitment mechanism. I forego the first episode to bind myself not to watch another. The cost of precommitment, however, only seems to buy me something I could have had for free, had I only been able to resist the temptation. We may consequently want to provide an argument that claims that it is rational to act against one's preferences at the time of temptation because doing so leaves one ultimately better off.

Two types of such arguments can be distinguished. On the first, it is sometimes rational to resist temptation because doing so is called for by the best deliberative strategy, by the agent's own lights. According to this type of argument, usually referred to as a *two-tier argument*, the rationality of the individual action should be assessed by whether it is endorsed by the best deliberative strategy. On the second type of argument, resisting temptation is rational because it is the product of mutually beneficial cooperation between the agent's time slices. Let us call these *time-slice cooperation arguments*.

In the following, I want to show that the instrumentalist arguments in favor of resisting temptation either fail or are redundant, save for a special case. To see why, we have to focus our attention on the question of what we take to be the standard of instrumental rationality when making these arguments. That is, what do we take to be the conative attitude by which we evaluate the agent's actions or deliberative strategies?

1.3   Preference-Based Instrumental Rationality

Instrumental rationality is traditionally understood as requiring agents to take the best means to ends they desire. But note that ends and desires do not appear in standard rational choice theory. Nor did they feature in the description of our temptation problem. Instead, standard rational choice theory, as well as much of the wider literature on practical rationality, features binary preferences. Given this ubiquity of preferences, how should we then think about the requirements of instrumental rationality?

On a broad understanding of instrumental rationality, actions or principles of choice are evaluated in light of the agent's own conative attitudes, or *proattitudes*.[3] If we adopt such a broad understanding, there is then an open question as to which of the agent's conative attitudes should be the basis of evaluation of the agent's actions. Rational choice theorists typically assume that this basis of evaluation should be the agent's preferences over the objects of choice, which, in the case of choice under certainty, are

---

[3] Williams (1979) seems to articulate such a broad understanding of instrumental rationality when he argues that an agent only has a reason to do *x* if doing *x* somehow advances an element in his "subjective motivational set" *S*. This subjective motivational set, according to Williams, could contain various different proattitudes, plans, or commitments.

outcomes. According to what, in the following, I want to call *preference-based instrumental rationality*, instrumental rationality is about acting well in the light of one's preferences over outcomes. That is, preferences form the standard of instrumental rationality. Outcomes, in turn, then play the role of ends in preference-based instrumental rationality. This notion of instrumental rationality requires that we take preference to be a binary kind of conative attitude, which matches the intuitive sense of preference we have been using so far.[4]

The move to a preference-based notion of instrumental rationality is very common but often implicit and seldom argued for.[5] Crucially for us, preference-based instrumental rationality appears to justify a requirement to maximize with regard to one's preferences, and thus sophistication, instrumentally. If instrumental rationality requires us to act well in light of our preferences over outcomes, then, provided that there is a most highly ranked outcome, instrumental rationality seems to require us to take the action that leads to it. If I choose in this way, I will not frustrate any of my binary preferences. Preference-based instrumental rationality thus seems to lend support to one side of the puzzle we started out with: if resisting temptation requires us to act against our preferences, then that seems instrumentally irrational. In the following, I will consider whether instrumentalist arguments in favor of resisting temptation, and thus of acting counterpreferentially, nevertheless go through under preference-based instrumental rationality.

1.4 Two-Tier Arguments

Two-tier arguments proceed from the observation that agents sometimes serve their ends best if they do not, at every point in time, take their reasons

---

[4] Preference is sometimes also interpreted behaviorally as a kind of disposition to choose, in particular by economists. Doing so would require us to look for the standard of instrumental rationality elsewhere, as Section 1.8 does.

[5] Many authors use *desire* and *preference* interchangeably. Others equate ends with outcomes. In this passage, for instance, Morris and Ripstein (2001: 1) claim that rational choice theory requires agents to have rankings of ends: "[t]he traditional theory of rational choice begins with a series of simple and compelling ideas. One acts rationally insofar as one acts effectively to achieve one's ends given one's beliefs. In order to do so, those ends and beliefs must satisfy certain simple and plausible conditions: For instance, the rational agent's ends must be ordered in a ranking that is both complete and transitive." Yet others claim that ends and desires are different from preferences over outcomes but still abide by preference-based instrumental rationality. Gauthier (1987: 22–26) claims that ends may be inferred from preferences but that preferences are primary and rationality is about maximizing a measure of preference. Nozick (1993: 144), too, claims that preferences are basic and that ends and desires can be derived from them through some process of filtering or processing. Hampton (1994) provides a critique of standard rational choice theory that relies on interpreting the theory in terms of preference-based instrumental rationality.

directly from their ends. Or, in terms of preference-based instrumental rationality, agents sometimes serve their preferences best if they do not act in accordance with their preferences at every point in time. This is the basic insight David Gauthier (1994) provides in his "Assure and Threaten." Given this basic insight, Gauthier argues that instrumental rationality in fact demands that we assess not individual choices but entire deliberative procedures by how well they serve our preferences. We then regard actions as rational if and only if they are in accordance with the best deliberative procedure – even if that procedure calls for a choice that serves the agent's preferences at the time of action less well than another.

There are various worries about the two-tiered nature of this account. For instance, we do seem to have a strong intuition that whether an action is instrumentally rational depends on how well it serves the agent's preferences at the time of action. Bratman (1998) calls this the *standard view.* Denying it would suggest that we can be moved by the "dead hand of the past," that is, by past preferences or by plans previously made. But what preferences I once held but no longer hold does not seem instrumentally relevant at the time of action. Neither do plans previously made that do not serve my current preferences. Under preference-based instrumental rationality, these considerations seem to support sophistication.

However, here I want to raise another, more fundamental problem for two-tier arguments for the rationality of resisting temptation. And that is that in temptation cases we cannot in fact establish that a deliberative strategy that endorses resisting temptation really serves the agent's preferences best. And so, under the assumption of preference-based instrumental rationality, the argument does not get off the ground. While this is a problem for two-tier accounts in general, let me first look at Gauthier's own in more detail.

Gauthier (1994) appeals to the counterfactual consideration that the agent at each point in time thinks that my current preferences. Under preference-based instrumental is better off going through with a resolution than she would have been had she made no resolution at all. The example that originally motivated Gauthier's argument is an intertemporal prisoner's dilemma between two agents first described by Hume (2007/1739: III.2.5 520–21). In this example, two farmers *A* and *B* would benefit from helping each other harvest their crops rather than doing it each on their own. However, for each, it would be even better if he received help with harvesting his field, without having to reciprocate. Now we imagine that the dynamic structure of this case, illustrated in Figure 1.2, is such that *A*'s field is ready to harvest earlier.

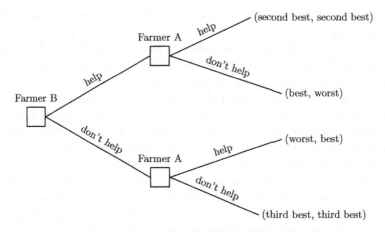

Figure 1.2 Intertemporal prisoner's dilemma

In this dynamic choice problem, if *A* maximizes with regard to his preferences once it is his turn, then whether he was himself helped or not, he will decide not to help *B*. Either way, he will be better off not helping. But knowing this, *B* will not help *A* in the first place. He would know that the favor would not be returned and so is better off not helping. The farmers thus each end up with a worse outcome than they could have had: they each end up harvesting their fields alone, when they could have helped each other and both been better off.

The farmers could achieve the better outcome, Gauthier argues, if *A* could make a sincere assurance to reciprocate to *B*, provided that *B* helps him first. If *B* believes this, he will in fact help, in order to secure *A*'s help in return. But, according to Gauthier, *A* can only make a sincere assurance that *B* will believe if he will take himself to have reason to follow through on the assurance when it comes to doing so. The problem here is that he will not take himself to have such reason if he takes his reasons directly from his preferences over outcomes at the moment of choice. In this kind of case, then, *A* does better with regard to his preferences if he uses a deliberative procedure that requires of him to go through with his assurance, even if it means at times not choosing the act that he prefers.

In a nutshell, the specific deliberative strategy Gauthier defends in this kind of case is the following. When it comes to acting on an assurance, the agent should ask himself two questions: (1) "how well would I have done if I had never made any assurance?" and (2) "how well will I do if I act on the assurance?" If the agent judges that he would do worse acting on the assurance than he would have done having never made one, then he is free to just maximize with regard to his preferences at each point in time. Otherwise, he should act on the assurance.

Let us assume that in the absence of an assurance, both agents are sophisticated and know this about each other. They thus know that in the absence of an assurance, they will each have to harvest their fields alone. Then Gauthier's deliberative strategy leads to the desired result in the intertemporal prisoner's dilemma, provided that A has made the assurance. When it comes to helping B, A judges that he will do better acting on his assurance than he would have done never having made an assurance at all.

Interestingly, if we simply substitute *resolution* for *assurance*, this deliberative strategy can also be used to justify going through with resolutions to resist temptations in the kind of case presented earlier. Suppose that the agent made a resolution to watch only one episode and then stop. At t_2, when it comes to following through, she considers what would have happened had she not made this resolution. Again, we assume that in the absence of a resolution, the agent is sophisticated. She would then not even have watched the first episode. This means, according to the agent's preferences at t_2, that she would have done worse not having made a resolution than she would do acting on the resolution. Even then, she prefers only watching one episode to watching none. And then, according to Gauthier's deliberative strategy, she should follow through with the resolution.[6]

1.5   The Failure of Two-Tier Arguments

There is, however, a crucial difference between the temptation cases and the intertemporal prisoner's dilemma, and I want to argue that this shows that Gauthier's argument is unsuccessful in the case of temptation, even if it were successful in the case of the intertemporal prisoner's dilemma.

---

[6] While Gauthier (1996) argues in favor of extending the two-tier account to justify resolution in temptation cases, Gauthier (1997) in fact expresses some skepticism about this. The motivation Gauthier (1997) offers for not extending the account to temptation is that the agent does not relate to himself over time as he does to other people. Part of the reason in favor of cooperation, Gauthier here claims, is that the agent views other people as "ends in themselves." But, according to Gauthier, the agent does not view his previous selves in that way. Gauthier here seems to abandon our presupposition and the presupposition he makes in "Assure and Threaten," that resolution is to be justified in terms of instrumental rationality alone. And so, while I agree in the following that temptation cases are crucially different from the intertemporal prisoner's dilemma, the difference Gauthier himself points out is one that he cannot appeal to if he wants to offer a true instrumentalist two-tier account.

In the intertemporal prisoner's dilemma, each farmer's preferences over the possible outcomes of the game remain constant. These constant preferences can be used as instrumental standards by which to evaluate the deliberative strategy Gauthier proposes. The temptation cases are different in this respect. It is in fact a defining feature of temptation cases as we characterized them that the agent does not have constant preferences over outcomes. Under preference-based instrumental rationality, these same changing preferences form the standard of instrumental rationality.

For a two-tier account to apply, we need to identify a deliberative strategy that is best by the agent's own lights. The deliberative procedure Gauthier proposes results in the best outcome according to the agent's preferences at t_1, but it does not lead to the best outcome according to the agent's preferences at t_2. At t_1, the agent thinks that the best course of action is one whereby she watches only one episode and then stops. But at t_2, according to her preferences, the best course of action for the whole choice problem is the one where she watches the first episode and then goes on to watch another. According to the agent's preferences at t_2, a deliberative procedure that endorses this course of action would be best.

Gauthier's proposed deliberative strategy can endorse making a resolution to not watch a second episode and going through with it, as we have seen. But it would not equally endorse making a resolution to watch two episodes and going through with that resolution – which is the best course of action according to the agent at t_2. At t_2, the agent would have no problem going through with such a resolution, of course. But at t_1, the agent takes it to be better to have made no resolution at all than to act in accordance with it and watch the first episode. This is because, at t_1, she prefers watching no TV over watching two episodes.

We can imagine possible alternative deliberative strategies, however, that would allow the agent to make a resolution to watch both episodes and go through with that resolution. For instance, the agent could adopt a deliberative strategy according to which, whenever the agent knows that she will later prefer to give into a temptation, she should plan to both face and give into the temptation all along. Of course, such a deliberative strategy requires the agent at t_1 to act contrary to what her preferences over outcomes are then: at that point she knowingly puts in motion the course of action leading to her least preferred outcome. However, Gauthier's proposed deliberative strategy also requires counterpreferential choice, namely at t_2, when it asks the agent to resist temptation. And so defenders of a two-tier argument cannot rule out this alternative deliberative procedure on that basis.

Given that the agent at t_2 would prefer a deliberative strategy that allows her to both face and give into temptation, Gauthier's proposed deliberative strategy is not the best deliberative strategy according to the agent's preferences at each point in time. It is the best deliberative strategy according to the agent's preferences at t_1, but it is not the best deliberative strategy according to the agent's preferences at t_2. Therefore, an argument that requires an agent's deliberative strategy to be best by her own lights in order for it to be rational to follow it does not go through. At the time when the agent is tempted, she does not think that a deliberative strategy that requires her to resist temptation is best. And so, according to such an argument, she would not be rationally required to follow it.

Now we may wish to respond that some deliberative strategies are simply not feasible for an agent and that two-tier arguments can only require agents to act in accordance with the best deliberative procedure out of the feasible set. For instance, we might suggest that a policy of both facing temptation and giving into it is one that the agent at t_1 would never adopt. If the agent at t_2 knows this, she might take that deliberative strategy to be simply infeasible and instead consider what is the best deliberative strategy out of the feasible set – perhaps the one proposed by Gauthier.

Again, this fails to support the two-tier argument in favor of resisting temptation. Suppose that we grant that the agent at t_1 will never adopt a deliberative strategy that is worse than some other available deliberative strategy with regard to her preferences then. By the same reasoning, we should grant that the agent at t_2 will never follow through with a deliberative strategy that leaves her worse off than some other available alternative strategy with regard to her preferences then. By the time the agent gets to t_2 and faces temptation, though, it is possible for her to switch to a deliberative procedure that allows her to give into temptation. This would be better with regard to her t_2 preferences than sticking to Gauthier's proposed strategy. And so the same reasoning that was supposed to rule out the strategy of facing and giving into temptation would also rule out rationally facing and resisting temptation in the way Gauthier proposes and thus undermine the two-tier argument. Again, the root of the problem is that in temptation cases, according to preference-based instrumental rationality, there is no stable standard against which to evaluate the agent's choice behavior.[7]

---

[7] Thanks to an anonymous referee for pointing out this potential objection and to José Bermúdez for suggesting the response.

Gauthier (1997) proposes a different deliberative strategy specifically for the context of temptation. There he notes that often, in cases of temptation, while the agent's proximate preferences, e.g., for watching a second episode, change, the agent retains "vanishing-point preferences" that still favor watching only one episode. These vanishing-point preferences are preferences about how to choose in similar situations in the future. So, even while the agent is tempted, she may prefer not to give into a similar temptation at future points in time. Let us grant that this is so in our TV consumption case. Even as I am tempted to watch another episode, I prefer that I only watch one episode at my coffee break the next day. Gauthier thinks that this makes it the case that the best deliberative strategy is one whereby the agent ignores her proximate preferences but acts in accordance with the vanishing-point preferences she holds at other times.

It is clear that given her vanishing-point preferences, the tempted agent judges that she will do much better by adopting a deliberative strategy that will make her resist temptation at all points in time than she would do if she adopted a deliberative strategy whereby she always gives into temptation. However, that does not make it the case that the agent takes the deliberative strategy of always going with her vanishing-point preferences to be *best*. In particular, a deliberative strategy whereby she can make just this one exception, to both face and give into this one temptation, would be preferred by the tempted agent. Gauthier's argument only goes through on the assumption that the agent is committed to adopting deliberative strategies that treat similar decision problems alike. However, such a commitment is not required by instrumental rationality. Without any desire for such consistency, the agent could always formulate deliberative procedures that allow for exceptions that are indexed to a specific time or place.

At this point, we might want to make a two-tier argument at a higher level, to the effect that agents who don't allow themselves to make exceptions generally do better in life. But again, as long as the agent's shifted preferences are the standard of instrumental rationality, the best deliberative strategy at this higher level will be one that allows just this one exception to not making exceptions. The underlying problem for both of Gauthier's accounts is that given preference-based instrumental rationality, as preferences change, the standard by which to evaluate deliberative strategies changes.[8] This is in fact fatal for any two-tier account. According

---

[8] I am assuming here that preference-based instrumental rationality is about doing well by the preferences the agent actually holds at the time of action. It is not, e.g., about doing well by all the preferences the agent has ever held, or will ever hold, or about doing well by the preferences the agent has held and will hold within some smaller window of time. By doing so, I am rejecting a temporally

to two-tier accounts, an action is rational if and only if it is endorsed by the best deliberative procedure. This approach shields the tempted agent's actions from being evaluated in terms of her shifted preferences directly. However, given preference-based instrumental rationality, the shifted preferences reappear at the higher level of deliberative strategies. And at the time of temptation, the agent is not only tempted but also would endorse a deliberative procedure whereby she would give into temptation.

1.6 Time-Slice Cooperation Arguments

Edward McClennen (1998) offers a treatment of temptation cases that is more explicit about the changing nature of the agent's preferences, which makes it impossible for us to judge the benefits of a deliberative procedure against a single set of preferences. He still thinks an appropriate, unchanging instrumental standard for this context can be formulated, however. His instrumentalist argument is based on intertemporal, intrapersonal optimality instead, a standard he had already advocated in McClennen (1990) in a slightly different context.[9]

At first sight, his account may look like another two-tier account. The deliberative strategy McClennen defends as rationally called for under many circumstances is resolution. Let a plan be a set of choices, one for each decision node the agent could find herself at in a given decision tree. Under certainty, each plan has one outcome associated with it. A resolute agent considers which plan or plans she prefers most at the outset, adopts one, and then simply carries it out. McClennen thinks that there are instrumental advantages to resolution whenever it makes possible a series of choices that is judged at least as good or better by the agent at each point in time in the decision problem than the alternative where she is sophisticated. That is, resolution can be justified by appealing to what we may think of as Pareto improvements between an agent's "time slices": resolution leaves some time slices better off and no time slice worse off.

Resolution in the above-mentioned temptation cases indeed yields such an intrapersonal Pareto improvement. If the agent makes a resolution to only watch one episode and does not give into temptation, she ends up with O_1. If, instead, she is sophisticated and acts according to her

extended view of the agent's interests. I do so for the same reasons as I reject interpreting McClennen's appeal to Pareto optimality between time slices as a two-tier account later.

[9] Intrapersonal optimality had also already been discussed in the economic literature as a choice criterion for agents with changing preferences. See Peleg and Yaari (1973).

preferences at each point in time, she ends up with O_0, as we saw earlier. But at each point in time in the dynamic choice problem, she prefers O_1 to O_0. So the resolute strategy is superior according to McClennen's criterion. And, in fact, no further Pareto improvements are possible here because there is no other outcome that is judged better by the agent at each point in time.

There is a substantive reason and a reason of argumentative strategy for not interpreting McClennen's argument as a two-tier argument. The substantive reason is that intrapersonal optimality is implausible as a standard of instrumental rationality. This is because an agent need not care about her preferences at different points in time. But treating intertemporal optimality as a standard of instrumental rationality would make it nonoptional for such an agent to cater to her past and future preferences. A requirement to cater to one's past or future preferences even if one does not care about them does not sound like a requirement of *instrumental* rationality (even if it may be a noninstrumental requirement of rationality). Instrumental rationality, I take it, is about doing well by the ends we actually hold at the time of decision.[10] Like Gauthier, McClennen himself claims to be in the business of establishing requirements of instrumental rationality. Under preference-based instrumental rationality as we understand it, if the agent did care about achieving intertemporal optimality in a way that is relevant for instrumental rationality, she would have ranked the Pareto-optimal outcome most highly in her preferences. Given that we postulated that the agent does not rank resisting temptation most highly at the time of temptation, she thus does not sufficiently care about achieving intrapersonal, intertemporal Pareto optimality.

The strategic reason for not interpreting McClennen's argument as a two-tier argument is that if we do so, everything I say later about abandoning preference-based instrumental rationality will apply to his argument thus understood. If instrumental rationality is also about catering to one's past and future preferences, then, if an agent's temporary preferences are in tension with such an intertemporal standard, as they are in temptation cases, they misrepresent the true standard of instrumental

---

[10] It is sometimes assumed in the decision-theoretic literature that choosing rationally consists of choosing well for your future self. Jeffrey (1965/1983) appeals to this idea when arguing for his version of evidential decision theory. Briggs (2010) uses it to analyze various decision-theoretic paradoxes. And L. A. Paul (2015) presupposes this when she argues that rational choice is impossible when we can't know what our future attitudes will be. Of course, most of us care to some extent how we will view our decisions in the future. But this is rarely all that matters for us, and it may matter to us in different ways.

rationality – in which case we do not need a two-tier argument to explain why it could be instrumentally rational to resist temptation.

McClennen's appeal to optimality is thus not best understood as part of a two-tier argument. Instead, the best way to interpret his appeal to optimality is in analogy with the role of Pareto optimality in interpersonal choice problems such as the prisoner's dilemma. In those games, there is no agent whose end it is to achieve Pareto improvements. It is simply the case that achieving a Pareto improvement serves both agents' preferences. This provides the basis for authors like Gauthier to argue for the rationality of decision rules that make cooperation possible. Each agent has a reason to do her part in making cooperation possible, because each agent stands to gain from it. McClennen suggests that, analogously, in the temptation cases the agent's time slices can engage in mutually beneficial cooperation.[11] Adopting a choice rule that makes such cooperation possible is advantageous for each time slice.

1.7   The Failure of Intrapersonal Optimality Arguments

Regardless of the merits of the argument in the interpersonal case, this analogy ultimately fails. McClennen leaves it vague what time slices are and how they relate to the agent. However we think of them, though, the analogy to interpersonal cooperation is suspect.

At one end of the spectrum, we could think of the time slices as separate agents that exist in succession (but presumably retaining memory of resolutions made by earlier time slices). Carrying out a resolute choice strategy now requires different agents to do their part: one needs to form a resolution, and the other ones need to carry it out. The problem with this interpretation, apart from the implausible picture of agency it paints, is that the time slice at t_2, whose turn it is to resist the temptation, is asked to act on a resolution that she did not make herself. She never made any assurance to the time slice at t_1 that she would resist the temptation and had no say in the formation of the resolution. She could not have done so because she was not around at the earlier points in time.[12]

---

[11] Ainslie (1992) similarly suggests that willpower in the face of the preference reversals caused by hyperbolic discounting is the result of a kind of intrapersonal cooperation.

[12] Bratman (1996) objects to appeals to intrapersonal optimality on the basis that the earlier time slice is not around anymore once the later time slice makes a choice. The concept of cooperating with the dead, as it were, seems odd. I take this objection not to be entirely decisive. Gauthier's proposed deliberative strategy, at least, does recommend going through with an assurance to the dead.

And so, if this case resembles a case of interpersonal cooperation, it resembles one where a cooperative scheme is forced on an agent. In the farmer case, suppose that farmers A and B have not communicated at all. Farmer A harvests half his field alone and then takes a break. When he comes back, farmer B has harvested the rest of the field for him. Even if A knows that B would only have done this had he expected A to return the favor, it does not seem instrumentally irrational of A not to return the favor. It might be nice to do so or even called for by some social norms, but unless A cares about these social norms or about being nice, instrumental rationality seems in fact to require A not to help his neighbor in return.

At the other end of the spectrum, we could think of time slices as different stages of the same agent. In the temptation cases, this same agent merely changes her preferences over time. But in this kind of case, we usually simply assume that when an agent changes her preferences, she changes her mind, and the new preferences simply override the old preferences. In this case, there is no reason for the agent to still act on preferences she does not hold anymore. If the agent is cooperating with herself in temptation problems, as we are supposing, then she is in fact cooperating with an agent who has changed her mind about the terms of cooperation. In interpersonal cooperation, at least, there appears to be no reason to make good on an assurance to her cooperator if doing so would not benefit her anymore due to a shift in her preferences.

For instance, in the farmer case, suppose that farmer A secured farmer B's help with an assurance, but just before it comes to reciprocating, B changes his preferences such that he now prefers harvesting alone after all. Perhaps he took a sudden dislike to A. It seems implausible that in this kind of case there is anything to be said for A helping B. In fact, it would be bizarre for A to impose his help against B's will. Likewise, it seems, in the case where the tempted agent is cooperating with herself, that there is nothing to be said for catering to the agent's earlier preferences once they have changed.

The best way to think about time slices may lie somewhere in the middle. But two requirements would need to be met in order for the argument to resemble interpersonal cooperation. First, it would need to be the case that time slices are unified enough such that a later time slice recognizes a resolution made by an earlier time slice as her own. But they can't be so unified that the preferences of later time slices override the preferences of earlier time slices. I don't see how these two requirements could plausibly be met together.

1.8   Giving Up Preference-Based Instrumental Rationality

I have argued that the two most prominent kinds of instrumentalist arguments for the rationality of resisting temptations fail. Two-tier accounts fail because in temptation cases the standard by which to evaluate deliberative procedures shifts. And no plausible account of mutually beneficial cooperation between time slices of an agent can be given. However, my argument relied on the assumption of preference-based instrumental rationality. But this assumption may well be false. I now want to suggest that giving up preference-based instrumental rationality does not help those who want to make the instrumentalist arguments for resisting temptation we discussed.

If there is to be any hope of rational choice theory formulated in terms of preferences to serve as a theory of instrumental rationality, then preferences should at least normally or ideally stand in close relationship to the true standard of instrumental rationality. This would be so, for instance, if we understood preferences as conative attitudes that act as a summary representation of the agent's underlying desires and concerns that form the true standard of instrumental rationality. Or it would be so if preferences were understood as dispositions to choose that are ideally responsive to the agent's underlying desires and concerns as a whole. Giving up preference-based instrumental rationality opens up the possibility that preferences express or represent this true standard of instrumental rationality incorrectly or incompletely. We may in fact suspect that this is the more appropriate analysis of temptation cases: under the influence of some tempting situation, the agent's preferences shift to diverge from her underlying, true desires. As I want to argue here, however, conceding this does not help those who want to make the instrumentalist arguments we discussed.

Suppose that at any point in time only one unique preference ranking can accurately capture the true standard of instrumental rationality. We can then distinguish two exhaustive possibilities of how the tempted agent's shifted preferences relate to the true standard of instrumental rationality. First, whether the agent's actual preferences correctly represent her underlying desires or not, the preferences that would do so are not stable. That is, the underlying true standard of instrumental rationality in fact shifts significantly over time. In this case, all the problems we discussed in the foregoing still arise, and the instrumentalist arguments still fail to establish the rationality of resisting temptation. The second possibility is that the agent's underlying desires would in fact only be correctly expressed by a stable preference ranking. In this case, the underlying true standard of instrumental rationality is in fact stable.

Thinking of temptation cases along the lines of this second possibility may in fact be what explained the intuitive instrumental irrationality of giving into temptation all along.[13] The fact that the preference reversal in temptation cases is only temporary could be seen as evidence that tempted agents never stop having the goal of being temperate but are only momentarily confused about what they really want.[14] However, in this case, it seems like we don't need the instrumentalist arguments we have been considering in the foregoing anymore. What is instrumentally rational is to do well by one's underlying desires and concerns. If the true standard of instrumental rationality, all the way through, uniquely supports only watching one episode, then, even as the agent is tempted to watch another episode, instrumental rationality requires her not to do so. This is so for straightforward reasons. As long as the agent refrains from watching the second episode, however she manages to do so, she is instrumentally rational. Moreover, the instrumentalist puzzle we started out with easily resolves: agents are now at best instrumentally required to maximize with respect to their preferences if the preferences correctly capture their underlying desires – which we are supposing they don't in temptation problems.

If these are indeed the only two possibilities of how the tempted agent's shifted preferences can relate to the true standard of instrumental rationality in temptation problems, then there would be no use for the instrumentalist arguments we considered earlier. Either they fail, or they are redundant. I would, however, like to point to an interesting third possibility, where a two-tier argument may again be of use. This possibility may arise if we allow for nonuniqueness in the sense that several different preference rankings express the true standard of instrumental rationality equally well. Suppose, then, that (1) there is at least one stable preference ranking that would correctly capture the agent's underlying desires at every point in time, but (2) at any point in time it is also true that several different preference rankings would accurately capture the agent's underlying

---

[13] This is suggested, for instance, by S. Paul (2015), who claims that the stable, more long-term preferences an agent has before and after being tempted have a better claim to "speak for the agent" (even at the time when she is tempted). Gauthier's (1997) argument that the agent should act on her "vanishing-point preferences" may also in part have been motivated by this intuition.

[14] However, note, too, that there may be cases where the agent's momentary preferences have a better claim to accurately representing what she truly cares about. This could be so, for instance, for a woman requesting an epidural when in labor despite an earlier, well-informed resolution not to do so. See Andreou (2014) for this example. One advantage of the view described here is that it could explain why, in these kinds of cases, instrumental rationality may demand giving into "temptation."

desires. For instance, suppose that throughout, the agent's underlying desires underdetermine whether she should have the preferences she does in fact have at t_1 or those she has at t_2 – both are permissible. As a matter of fact, the tempted agent has shifting preferences. But she could have stable preferences that would capture the true standard of instrumental rationality correctly at every point in time. This general possibility may arise both when the standard of instrumental rationality is stable and when it shifts only slightly over time so that permissible preference rankings overlap.

In these circumstances, a two-tier argument may actually give the agent reason to stick with one of the permissible preference rankings throughout, or to act as if she did. Adopting a deliberative strategy that demands this kind of stability in the face of nonuniqueness will keep her from ending up with an outcome that is definitely worse according to her underlying desires, such as the outcome of not watching any TV. While this is an interesting possibility, it seems to me to be a special case, and only some real-life temptation cases will be accurately described by this analysis if nonuniqueness is even a coherent possibility. And, in any case, the kind of two-tier argument sketched here differs substantially from the ones typically presented in the literature on temptation cases.

1.9 Conclusions

If we stick to preference-based instrumental rationality, the instrumentalist arguments for resisting temptation fail. Only if we abandon it will we be able to give an instrumental argument for resisting temptations. There are in fact good independent reasons for abandoning preference-based instrumental rationality, at least if we have in mind preferences as they feature in standard rational choice theory. For one, in ordinary speech we often take conative attitudes over features of outcomes to explain our preferences over outcomes: I may prefer O_1 to O_2 because I desire to get my work done, and I take this to outweigh my desire to watch TV. The kind of instrumental failure that may be involved in temptation according to this picture also seems familiar. I often find myself forming all-things-considered attitudes over my options that on reflection did not do full justice to everything I care about in those options. And then I take myself to be instrumentally criticizable. Lastly, I argue elsewhere[15] that standard instrumentalist arguments in favor of the core requirements of rational choice theory do not work on the preference-based picture.

---

[15] See Thoma (2017).

Where does abandoning the preference-based picture leave us with respect to the rationality of resisting temptation? As I said earlier, if we can show that the agent's underlying concerns in fact are stable and that the agent's preferences merely momentarily misrepresent this fact, then resisting temptation is instrumentally rational for straightforward reasons. But this response depends on the true standard of instrumental rationality in fact being stable, which may or may not be true depending on the case. We can thus no longer give an argument that instrumental rationality *requires* that agents resist temptation as we characterized it. After all, instrumental rationality cannot demand that the agent have any particular ends. The best we can do, if we want to appeal to instrumental rationality alone, is to argue that agents ordinarily have desires that support resisting temptations in a wide variety of cases. Alternatively, we could make it a defining feature of temptation cases, properly understood, that the agent's underlying true desires in fact stably support not giving into temptation. If we do so, it must be clear that there is no more puzzle about how resisting temptation can be rational. The challenge then lies only in how agents can be motivated to do what is rational.

References

Ainslie, G. 1992. *Picoeconomics: The Strategic Interaction of Successive Motivational States within the Person.* Cambridge: Cambridge University Press.

Andreou, C. 2014. Temptation, resolutions, and regret. *Inquiry* 57(3):275–92.

Bratman, M. 1995. Planning and temptation. In *Mind and Morals: Essays on Ethics and Cognitive Science,* ed. Larry May, Marilyn Friedman, and Andy Clark (pp. 293–310). Cambridge, MA: MIT Press.

Bratman, M. 1998. Toxin, temptation, and the stability of intention. In *Rational Commitment and Social Justice: Essays for Gregory Kavka,* ed. Jules Coleman, Christopher Morris, and Gregory Kavka (pp. 59–83). Cambridge: Cambridge University Press.

Briggs, R. 2010. Decision-theoretic paradoxes as voting paradoxes. *Philosophical Review* 119(1):1–10.

Gauthier. D. 1987. *Morals by Agreement.* Oxford: Oxford University Press.

Gauthier, D. 1994. Assure and threaten. *Ethics* 104(4):690–721.

Gauthier, D. 1996. Commitment and choice. In *Ethics, Rationality, and Economic Behaviour,* ed. F. Farina, S. Vannucci, and F. Hahn (pp. 217–43). Oxford: Oxford University Press.

Gauthier, D. 1997. Resolute choice and rational deliberation: a critique and a defense. *Nous* 31(1):1–25.

Hampton, J. 1994. The failure of expected-utility theory as a theory of reason. *Economics and Philosophy* 10(2):195–242.

Holton, R. 2009. *Willing, Wanting, Waiting*. Oxford: Oxford University Press.

Hume, D. 2007/1739. *A Treatise of Human Nature*. London: Clarendon Press.

Jeffrey, R. 1965/1983. *The Logic of Decision* (2nd edn). Chicago, IL: University of Chicago Press.

McClennen, E. 1990. *Rationality and Dynamic Choice: Foundational Explorations*. Cambridge: Cambridge University Press.

McClennen, E. 1998. Rationality and rules. In *Modeling Rationality, Morality, and Evolution*, ed. Peter Danielson (pp. 13–400). Oxford: Oxford University Press.

Morris, C., and A. Ripstein. 2001. Practical reason and preference. In *Practical Rationality and Preference*, ed. Christopher Morris and Arthur Ripstein. Cambridge: Cambridge University Press.

Nozick, R. 1993. *The Nature of Rationality*. Princeton, NJ: Princeton University Press.

Paul, L. A. 2015. *Transformative Experience*. Oxford: Oxford University Press.

Paul, S. 2015. Doxastic self-control. *American Philosophical Quarterly* 52:145–58.

Peleg, B., and M. Yaari. 1973. On the existence of a consistent course of actions when tastes are changing. *Review of Economic Studies* 40(3):391–401.

Thoma, J. 2017. Advice for the steady: decision theory and the requirements of instrumental rationality. Ph.D. thesis, University of Toronto.

Williams, B. 1979. Internal and external reasons. In *Rational Action*, ed. Ross Harrison (pp. 101–13). Cambridge: Cambridge University Press.

CHAPTER 2

Self-Prediction and Self-Control

*Martin Peterson and Peter Vallentyne**

2.1 Introduction

The aim of this chapter is to examine the conditions of rationally permissible choice for agents that have fully rational preferences and beliefs, are fully aware of their dispositions to make choices under various conditions, but *may have only limited self-control over their future actions.* In the literature, there are three main conceptions of rational choice for agents facing a sequence of choice situations: (1) the resolute choice conception (e.g., McClennen 1990), according to which agents should adopt a plan with the best prospects and then simply comply with that plan at each choice situation, (2) the sophisticated choice conception (e.g., Schick 1986), according to which agents should predict how they will choose in future choice situations and then make a current choice that has the best prospects based on those predictions, and (3) the wise choice conception (e.g., Rabinowicz 1995), which is the same as the sophisticated conception except that it allows that conformance with a rationally adopted plan can count in favor of choosing an option, whereas the sophisticated choice conception does not. We endorse, develop, and defend the wise choice conception. We propose, controversially, that rational agents predict how they will choose in the future by ascribing subjective probabilities to their own future choices. They thus treat future choice nodes as a kind of chance event. We argue that backward induction is not, as commonly assumed, a necessary element of the sophisticated and wise choice conceptions. In our view, it is merely a method that rational agents may use, in special cases, to determine what to choose in light of their predictions about their future choices.

* For helpful comments, we thank Wlodek Rabinowicz, Johanna Thoma, and Paul Weirich.

2.2 Limited Self-Control

Our key points concern cases of *limited self-control* (*limited willpower*), which we understand broadly to include both (1) a *synchronic form* (*akrasia*), in which the agent chooses something, including resolutions or plans, that he judges to be rationally impermissible (e.g., not best), and (2) a *diachronic form* (*lack of resoluteness*), in which the agent lacks the full ability to determine, by virtue of a resolution (plan adoption), what choices he will make in future choice situations if he reaches them.[1]

Imperfect *synchronic* self-control (*akrasia*) arises when an agent's *motivational preferences*, which determine his actual choice disposition, are not aligned with his *rational preferences*, which reflect the agent's values (i.e., normative considerations that, for him, speak for or against the various options). Imperfect *diachronic* self-control (lack of resoluteness) arises when an agent does not always choose in accordance with previously adopted plans. This arises, for example, when an agent's future motivational preferences do not give lexical primacy to conformance with adopted plans. If an agent's rational preferences (values) give lexical primacy to such conformance, and if the agent has perfect synchronic self-control (no *akrasia*) now or in the future, then his future motivational preferences will also give lexical primacy to such conformance, and the agent will be perfectly resolute.

To illustrate the issue, consider the following example from Carlson (2003). Suppose that Alice knows that she will get two opportunities for eating a chocolate bar. Suppose further that the best outcome for her is eating a chocolate bar on the first occasion but not on the second (because eating one bar earlier is better than eating one bar later). The second-best outcome is eating a chocolate bar on the second occasion but not on the first (because one bar is better than none or two). The third-best outcome is eating no chocolate bar (because being hungry is better than eating two bars). The worst outcome is eating a chocolate bar on both occasions (because it is very unhealthy). Suppose that Alice knows all this. Is it rationally permissible for Alice to eat the chocolate bar on the first occasion?

---

[1] Holton (2009) understands self-control and willpower narrowly to involve only the diachronic sense (resoluteness) and not the synchronic sense (lack of *akrasia*). In this chapter, we use the terms broadly to cover both and make no claim about what the most common usage of this term might be. Also, we understand *akrasia* as making a choice that the agent judges to be rationally impermissible (or not sufficiently good) relative to specified values of the agent (moral, prudential, or all relevant values). Standard conceptions of *akrasia* appeal to all relevant values of the agent (practical reasons).

If Alice does not suffer from *akrasia* and is fully aware of her choice dispositions, it is, we claim, rationally required that she choose to eat the chocolate bar on the first occasion (the best outcome). If, however, she suffers from later *akrasia* and has no resoluteness, and she will eat a second chocolate bar in any case, then it seems plausible that if she knows this, it is rationally impermissible for her to eat a bar on the first occasion. Finally, if she suffers from *akrasia* but is perfectly resolute, then it seems plausible that it is rationally required that she adopt a plan not to eat the second chocolate bar, eat the first chocolate bar, and then comply with her plan. Or so we shall argue.

2.3   Parametric Dynamic Rational Choice

We focus on what rational choice requires in the context of parametric dynamic choice situations under risk. These dynamic choice situations are *parametric* in that the outcomes of an agent's choices depend solely on the choices he makes and on chance events (acts of nature). Unlike *strategic* choice situations, outcomes do not depend on choices made by other agents. We focus on *dynamic* choice situations, which are situations in which the agent makes a sequence of choices over time (not merely a single choice). We focus on choice *under risk*, where we assume that the agent assigns probabilities (and not merely possibilities) to chance events. We explore the extent to which rational choice in such contexts is also based on the probabilities the agent assigns to his future choices.

Our aim is to determine what (feasible) options, in any given choice situation, are *rationally permissible*. If there is only one rationally permissible option, then it is *rationally required*. If there are several, then each is *rationally optional*. We shall assume that a choice is rationally permissible if and only if it is a *best feasible option*, i.e., one that is at least as good (relative to the agent's preferences) as any other feasible option. Our core argument can be generalized to cover satisficing theories and to cover cases where the agent's preferences are incomplete (in which case, rational choice requires a feasible option that is not worse than other feasible options). For simplicity, we assume a maximizing conception of rationality with complete preferences.

As noted earlier, we shall consider three conceptions of rationality: resolute choice, sophisticated choice, and wise choice.[2] In order to introduce these views carefully, we first need to say more about parametric dynamic choice.

---

[2] We shall not discuss the myopic conception of rational choice, since it is rejected by all.

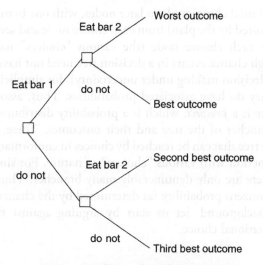

Figure 2.1 Chocolate: deterministic case

A parametric dynamic choice situation is a choice situation in which there is only one agent, who faces a sequence of choice situations, each of which involves a set of feasible choices for the agent. This can be represented by a decision tree (see, e.g., McClennen 1990: chap. 6), which consists of branching choice nodes (represented by squares) and branching chance nodes (represented by circles). Each branch from a choice node represents a choice the agent might there make, and each branch from a chance node represents a "choice" that "nature" might there make. If the "choices" of nature have probabilities, they are specified as part of the decision tree. Throughout, we shall assume, as is standard, that there is an initial (choice or chance) node to which all else is connected.

Our chocolate example, in which there are no chance events, can be represented by a decision tree (Figure 2.1). In dynamic choice situations, agents may adopt *contingency plans* (hereafter *plans*), which specify a choice for each choice node reachable from the initial node by a sequence of choices specified by the plan in question and by chance events.[3] If there are no chance events, a plan is simply a branch from the initial node through various choice nodes. If there are chance events, a plan will be a subtree

[3] Contingency plans do not specify how to choose, should the agent fail to comply with the plan at earlier choice nodes. A more general approach would focus on *strategies*, which tell the agent how to choose at every possible choice. For simplicity, we focus on contingency plans.

connecting an initial choice node to later nodes, with one branch segment
(the choice required by the plan) from each choice node and several branch
segments from each chance node (the various "choices" nature might
make). Although chance events in a decision tree need not have associated
probabilities (decision making under uncertainty), for simplicity, we shall
assume that they do have associated probabilities. Thus, associated with
each plan, there is a *prospect*, which is a probability distribution over the
various full branches of the tree and their outcomes. These are the full
branches of the tree that can be reached by choices in conformance with the
plan, in conjunction with various "choices" of nature. For simplicity, we
assume that there are only denumerably many branches. Thus each such
branch has a nonzero probability (as determined by the chance nodes).

    With this background, let us start by arguing against the resolute
conception of rational choice.

2.4 Resolute Choice

According to the *resolute* choice conception, rationality requires agents to
comply with the plans (or resolutions) they have adopted in the past. More
exactly, it does not require you to be "unconditionally committed to
execute a chosen plan." It merely requires that "if on the basis of your
preference for outcomes, you [rationally] adopt a given plan, and if
unfolding events, including any conditioning circumstances, are as you
had expected them to be, then you proceed to execute that plan"
(McClennen 1990, 1997: 232). That is, rational choice does not require
unconditional resoluteness. It only requires *rational resoluteness*, which is
a kind of conditional resoluteness. We understand this to be the disposition to comply with adopted plans when (1) it was rationally permissible to
adopt the plan at the time of adoption, and (2) the agent has acquired no
new unanticipated information that, if available to the agent at the time of
the plan's adoption, would have undermined the rationality of adopting
that plan. We here understand the second clause to be violated when the
agent is aware that he has failed to comply with the previously adopted
plan.

    It's worth noting that perfect resoluteness (unconditional or rational) is
compatible with *akrasia* (choosing something that one judges to be rationally impermissible). An agent can akratically adopt a rationally impermissible plan and still be perfectly resolute in her implementation of the plan.
Indeed, perfect *unconditional* resoluteness requires such implementation,
whereas perfect *rational* resoluteness is silent about such cases.

Although McClennen is not explicit about this, we assume that the resolute conception of rationality also requires agents to *adopt* plans whenever adoption is rationally required. If plans are never rationally adopted, the requirement to comply within rationally adopted plans never has any force. This is relevant because the resolute theory of rational choice appeals to plans that were *actually rationally adopted* and not to plans that it *would have been rational to adopt*. In this respect, the theory is like actual consent theories of political obligation and unlike hypothetical consent theories thereof.

When, then, is it rational to adopt a given plan? The resolute conception of rational choice holds that it is rationally permissible to adopt a given feasible plan if and only if *full implementation* of the plan has prospects that are at least as good as those of any alternative feasible plan. We use feasibility in an objective sense and assume that the set of feasible options (abstractly possible choices for the agent in the situation) is the set of options that the agent could choose if he were fully informed and had a suitable choice disposition. The resolute conception thus requires that one comply with any plan adopted if (1) at the time of adoption, full implementation of that plan has, relative to the agent's evidence, prospects that are at least as good as those of full implementation of any alternative feasible plan, and (2) since then, there has been no unanticipated new information that undermines the previous condition.[4]

We have two objections to this conception of rational choice. First, the rational permissibility of plan adoption is, we claim, based on a *realistic* assessment of how likely the agent is to fully comply with the plan. Appealing to the idealization of full compliance abstracts from crucial features of the choice situation. Few, if any, agents are perfectly disposed to comply with plans that they adopt. In our chocolate example, the best plan based on an *idealized* (full compliance) assessment is to eat the chocolate on the first occasion but not on the second occasion. Realistically, however, this is the worst plan to adopt because it will not be fully executed and will lead to eating chocolate on both occasions (the worst outcome). We agree with McClennen that for agents with perfect self-control, it would be rationally permissible to be resolute. So we are not directly challenging McClennen's view about what plans should be adopted by agents with perfect self-control. Our point is that in the general

---

[4] McClennen (1990: sec. 12.6) further suggests that resoluteness is feasible only where all temporal selves (of the agent) can reasonably expected to benefit from resoluteness. For simplicity, we ignore this issue.

case, where agents may lack perfect self-control, this theory makes little sense. That is, our first objection is not that McClennen's theory is false but rather that its scope is too narrow. The resolute choice conception is not applicable to the types of cases we care about: realistic choice situations in which the agent may lack perfect self-control.

Second, even if an agent is perfectly disposed to comply with whatever plan is adopted, conditional on the adopted plan being a best plan to adopt in the realistic sense, it is not always rational to *comply* with such a plan. In the well-known toxin example (Kavka 1983), it may be that adopting the plan to drink the sickness-inducing toxin is the realistically best plan to adopt (because its adoption leads to a reward), but given that the reward is irrevocably given before drinking the toxin, it can be irrational to drink the toxin, if this is still feasible. The mere fact that it was rational to adopt the plan, and no unanticipated new information undermined the support for the plan, does not ensure that it is rational to comply with the plan.

It is important to note, however, that there is a certain kind of agent for which the resolute conception of rational choice gives correct answers. Suppose that (1) an agent has rational preferences that accord *lexical primacy* to conformance with adopted plans (or more weakly, when such plans have the best prospects under the assumption of full compliance) and (2) she suffers from no *akrasia* (i.e., her motivational preferences are always in accord with her rational preferences). Call such an agent *perfectly resolute in the evaluation-based sense* (because her resoluteness comes from her lexically primary rational preference to comply with adopted plans). The resolute conception of rationality gives the correct answers with respect to such an agent. For such an agent, a plan has the best prospects under realistic assumptions about compliance if and only if it has best prospects under idealized (perfect compliance) assumptions. Moreover, given her lexically primary rational preference for conformance with adopted plans, it is always better for the agent to comply with such plans than not to do so. So, for such an agent, the resolute conception of rational choice is correct. The problem, of course, is that no real agent is such an agent, and indeed, this is a very special kind of idealized agent.

McClennen (1997: 239–42) holds that agents can be perfectly resolute, at least when the adopted plans are rationally adopted and not later undermined, even without any preference for compliance with (e.g., rationally) adopted plans. A mere act of the will (e.g., a rational commitment) can make the agent perfectly disposed to comply with (e.g., rationally) adopted plans. Call such agents *perfectly resolute in the commitment-based sense*. We find this type of resoluteness mysterious and psychologically unrealistic. Nonetheless, for such agents, as for the corresponding evaluation-based agents, a plan has the best prospects under idealized (full-compliance) assumptions if and only if it has the best prospect under realistic assumptions. So the resolute conception of rational choice correctly identifies the best plan of such agents to adopt. The problem is that it also entails that compliance with such plans is always rationally required. Given, however, that the agent need have no rational preferences for such compliance, it will sometimes say that compliance is rationally required when some alternative feasible action has better prospects (e.g., drinking the toxin is required, even when one is able not to do so). So the resolute conception gives the wrong assessments even for perfectly resolute agents when their resoluteness is commitment based rather than evaluation based.

2.5 Historical Separability and Normal-Form/ Extensive-Form Coincidence

Before turning to the assessment of the sophisticated conception of rational choice, it will be useful to discuss briefly two main conditions at issue between the competing conceptions of rational choice, as identified by McClennen. One condition is *historical separability* (McClennen 1990: 122), which requires that the rational permissibility, at a given node, of a plan for the future *not depend* on what the past was like.[5] In formulating this condition, McClennen implicitly assumes that agents are fully informed of all facts. If, however, agents are less than fully informed, and if rational permissibility is relative to the agent's beliefs (or at least the beliefs support by her evidence), then the historical separability would need to be reformulated to say that rational permissibility does not depends on the agent's beliefs about the past. For simplicity, we focus on fully informed agents, and we thus ignore this complexity. The resolute conception violates this condition because it makes rational permissibility depend on what plans the agent has rationally adopted in the past.[6] Because we believe that rational preferences can be historical (e.g., one

---

[5] McClennen (1990) calls this condition "separability," but we label it "historical separability" in order to make explicit that the issue is separability *over times*, as opposed to over people or states of nature. Also, here and later we give intuitive versions of McClennen's conditions, but these are meant to capture McClennen's more formal formulations.

[6] A fully adequate decision-tree representation of plan adoption would have to include choice nodes for the choices to adopt plans, as well as "regular" choice nodes (see Rabinowicz [2017] for discussion). For simplicity, we ignore this. Also, note that for historically sensitive rational preferences, the ranking of branches will be node relative *and not fixed for the tree*. In our chocolate example, for example, the branch of eating a chocolate bar at t_2 is ranked higher than the branch of

can prefer eating a chocolate bar to not eating one when one has not recently eaten a bar but have the opposite preference if one has recently eaten a bar), we agree with McClennen that historical separability should be rejected. We further agree that there is nothing irrational about having a pro tanto preference for complying with nonundermined plans that were rationally adopted in the past. Indeed, we think that it need not be irrational (although it's strange) to have preferences that make compliance with previously adopted plans *lexically prior* to all other considerations. So we agree with McClennen that rational choice need not satisfy historical separability (although perhaps for different reasons).

A second condition discussed by McClennen is the *normal-form/exten-sive-form coincidence* condition (McClennen 1990: 115). This requires that, in dynamic choice situations, a plan is rationally permissible relative to a given choice node when represented in *extensive form* (i.e., as a sequence of choices, as represented in a decision tree) if and only if it is rationally permissible, relative to that node, when represented in *normal form* (i.e., as a *single choice* of feasible sequence of "choices"). As discussed earlier, this is plausible for agents that are *perfectly* resolute and confront no unanticipated undermining information, but (1) few, if any, agents are so resolute, and (2) the condition, we shall now argue, is implausible for agents that are not. Consequently, the condition is implausible as a general condition of rational choice.

A plausible theory of dynamic choice should take into account all the information available to the agent at each choice node, and this includes information about how likely she is to make various choices at future choice nodes. The normal-form/extensive-form equivalence condition precludes such sensitivity because it treats a sequence of choices as a single choice. For example, in our chocolate example, the best plan open to the akratic agent is to have chocolate on the first occasion but not on the second occasion. The resolute conception of rational choice then says that rationality requires adopting that plan and then complying with it. The agent will comply with the plan at the first choice (by eating chocolate), but given her *akrasia*, she will fail to comply with the plan at the second choice node, and the result will be the worst possible result. Rational choice requires sensitivity to *how the agent is disposed to choose* at the various later choice nodes, but the resolute conception is insensitive to such information for agents that are not perfectly resolute.

---

not eating one if the agent did not eat a bar a t_1, but the opposite ranking holds if she did then eat a bar.

The resolute conception of rational choice requires ignoring information about the agent's future choice dispositions. As we will show in following sections, this problem can be overcome by being sensitive to the probabilities of future choices.

2.6 Sophisticated Choice

On the sophisticated choice conception of rationality (e.g., Schick 1986), the agent should first predict her final choices and then reason backward for earlier choices. For example, if she predicts that, due to *akrasia*, she will eat the chocolate bar on the second occasion even if she eats the chocolate bar on the first occasion, she should then compare the outcome of not eating chocolate on the first occasion with the outcome of eating chocolate on the first occasion. Because the former (chocolate only on the second occasion) is, we are assuming, rationally preferable to the latter (chocolate on both occasions), rational choice requires her not to eat the chocolate on the first occasion. In this view, the agent predicts future choices and then deliberates about her current choices in light of those predictions. This seems roughly correct, but we shall now generalize this approach.

It is generally assumed (e.g., by Schick 1986; McClennen 1990; Rabinowicz 1995) that backward induction is a necessary element of the sophisticated choice conception, but this is not so. Backward induction is just a tool the sophisticated agent uses for *making decisions based on predictions about future choices*. In principle, any other tool that does the job can be used by the sophisticated agent. Moreover, backward induction does not always work. First, if, for some branches, there is no final choice node, then there is no starting point for the backward induction. Second, even if all branches have final choice nodes, if there is more than one option that the agent might choose, and these are not equally valuable (e.g., where a less valuable option might be irrationally chosen), then backward induction will not get started in this case either.

A more general approach for sophisticated agents is to assign *probabilities* to each of his choices in each future node. These probabilities define the *choice disposition* that the agent predicts he will have at that node. In the simple finite case where, at each choice node, there is only one choice that the agent might make with nonzero probability, backward induction will be possible, but it is not essential. Instead, associated with each possible choice in the current choice situation, there will be a *prospect*, that is, a probability distribution over branches. The probabilities of the branches will be determined by the probabilities at future *chance nodes* and by the probabilities at future *choice nodes*. Rationally permissible choice requires that an agent choose a feasible option with a best (or at least maximally good) prospect. Here we leave open what the criteria are for the goodness (relative to rational preferences) of prospects. In particular, we do not assume that they must be risk neutral or based on expected values. We only assume that prospects are assessed rationally, whatever that requires.

The sophisticated conception of rational choice is thus best understood as requiring that the agent choose an option with rationally best prospects, where prospects are probability distributions over branches, and where the probabilities reflect both the probabilities of future chance events and the probabilities of future choices by the agent.[7]

For agents that are perfectly rationally resolute in the evaluation-based sense, the sophisticated conception agrees with resolute conception, as noted earlier. If, however, the agent is not perfectly rationally resolute, then conforming to the nonundermined rationally adopted plan may not be rational (given that he may not later comply with it). In our chocolate example, suppose that the agent is sure to eat the chocolate on the second occasion, even if she adopts the plan not to do so (due to *akrasia*). In this case, the agent faces a choice between eating the chocolate on the first occasion with 100 percent chance of eating the chocolate at the later time and not eating the chocolate on the first occasions with 100 percent chance of eating the chocolate at the later time. Since the latter prospect is rationally better, the rational choice on the first occasion is to choose not to eat the chocolate. Here the resolute conception wrongly entails that the agent is rationally required to eat the chocolate on the first occasion.

Moreover, the sophisticated conception handles cases where the probabilities of future choices are not zero or one. Suppose that the agent is 50 percent likely to eat the chocolate on the second occasion if she eats the chocolate on the first occasion, and she is 100 percent likely to eat the chocolate on the second occasion if she does not eat the chocolate on the first occasion. Then the rational permissibility of eating the chocolate on the first occasion will depend on the how valuable the different outcomes are. Suppose, for example (Figure 2.2), that eating chocolate twice has a value of 4 and eating chocolate only on the second occasion has a value 10. Then, assuming for illustration that the value of prospects is their expected value, (1) if the value of eating the chocolate only on the first occasion is more than 16, then it will be rationally permissible to eat the chocolate on

---

[7] See McClennen (1990) for discussion of prospects and of probabilistic choice.

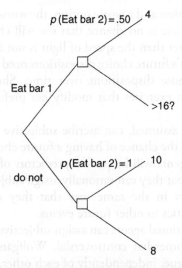

Figure 2.2 Chocolate: probabilistic case

the first occasion (with a 50 percent chance of also eating it on the second occasion), and (2) if that value is less than 16, then it will be rationally impermissible to do so.

The sophisticated conception of rational choice, as just sketched, is, we believe, essentially correct. A crucial assumption, however, is that agents can rationally assign probabilities to their future choices. In Section 2.7, we defend this assumption. Following that, we address the issue of whether the sophisticated conception is committed to historical separability (the irrelevance of the past).

2.7 Can Probabilities Be Ascribed to Future Choices?

To render the notion of self-predictive probabilities sharp, we distinguish between the *feasibility* of a plan (or a choice) and the objective *probability* that the agent will execute it. As explained in Section 2.4, the set of feasible options (abstractly possible choices for the agent in the situation) is the set of options that the agent could choose if she were fully informed and had a suitable choice disposition. An agent's *choice disposition* is the objective probability, for each feasible option, of the agent choosing that option. Feasible options can, in this sense, have zero probability if there is no chance of the agent choosing it (even if she could). Thus, for example, if a perfectly rational and fully informed agent, with no *akrasia*, has a choice between a better option and a worse option, the worse option is feasible for her, even though there is no chance that she will choose it.[8] In contrast, choosing to run faster than the speed of light is not feasible for any agent. Obviously, an agent's future choice dispositions need not be fixed. She may be able to alter those dispositions over time. She may, for example, choose to engage in exercises that modify her preferences or reduce her weakness of will.

Agents, we have assumed, can ascribe subjective probabilities to their future choices (e.g., the chance of having a future choice disposition). Our claim is not that agents will be perfect predictors of their future choices. It is rather simply that they can rationally assign subjective probabilities to their future choices in the same way that they can rationally assign subjective probabilities to other future events.

The claim that rational agents can assign subjective probabilities to their future choices is somewhat controversial. Wolfgang Spohn (1977) and Isaac Levi (1989) argue, independently of each other, that a rational agent cannot at the same time deliberate and predict his own behavior. As Levi puts it, "deliberation crowds out prediction." We grant that it may not be possible to deliberate about one's *current* choice while at the same time predicting what one's *current* choice will be. Fortunately, we do not require this. We hold that agents can, and should, *predict* their choices in *future choice situations* and *deliberate* about their choices in the *current choice situation*. (For insightful discussion of this issue, see Rabinowicz 2002.)

Still, there is a potential problem, at least if we assume that an agent's probabilities should be reflected in the bets that he is willing to accept. An agent who ascribes subjective probabilities to his future choices must be prepared to accept certain bets on which act he will eventually choose. By accepting such bets, however, it becomes more attractive for the agent to perform the act on which he is betting. Suppose, for instance, that you are disposed to choose, at a future choice node, A with probability 0.6 and B with probability 0.4 and that you predict that you will choose these options with these probabilities. This entails that you should be willing to accept a bet in which, if you reach this choice node, you win 1 unit of value if you choose A and lose $0.60/0.40 = 1.5$ units of value if you choose B. (The expected value of A is $1 \cdot .6 = 0.6$ and that of B is $-1.5 \cdot .4 = -0.6$).

---

[8] We assume that choices are mental acts and that the physical acts, such as pushing a button, are outcomes of mental acts. Thus, where (as is typical) agents have imperfect control of their physical actions, the chance of the physical action being executed, given that it is chosen, will be less than 1. We treat this simply as one of the many ways in which outcomes may not be fully determined by the choice of the agent.

Why is this a problem? Once you have accepted the bet described earlier, you have an incentive to make sure that you win the bet, if you reach the choice node. You can easily do this simply by choosing *A* if you reach the node. So once the bet is accepted, the probability of choosing *A*, should the choice node be reached, is 1.0. This is problematic because we initially assumed that the agent was disposed to choose *A* with probability 0.6. That is, by merely predicting his own choice, the agent's choice disposition increases from 0.6 to 1. This suggests that self-prediction, even for future choices, is problematic.

Spohn and Levi spell out the technical details of this argument in somewhat different ways, but the key idea in both accounts is that the bets we use for eliciting self-predictive subjective probabilities interfere with the entity being measured. The following analogy might be helpful: if you measure the temperature of a hot cup of coffee with a very cold thermometer, the temperature of the measurement instrument will significantly affect the temperature of the coffee. That is, your measurement instrument interferes with the entity being measured. The only way to make sure that this effect does not occur is to ensure in advance that the thermometer has exactly the same temperature as the coffee, but this is possible only if we know the temperature of the coffee before we measure, which makes the measurement process superfluous. According to a more radical, operationalist version of this argument, we should conclude that because we cannot measure the temperature of the coffee, the theoretical term *temperature* has no meaning.

When bets are used for measuring probabilistic choice dispositions, the bets are by no means superfluous, but they influence the entity being measured. So, although the analogy with hot coffee is not perfect, the underlying phenomenon is similar, irrespective of which version of the objection one prefers. (Levi seems to endorse the operationalist version, according to which the impossibility to measure self-predictive probabilities makes the concept meaningless.)

However, in opposition to Spohn and Levi, we think this measurement theoretical effect is not a reason for giving up the idea that rational agents can make probabilistic self-predictions. One might, of course, question whether betting ratios are the proper way of measuring degrees of belief, but here we shall not question this. We shall instead indicate how interference between accepting bets and predictive acts can be controlled. First, the stakes of the bets can be set to be negligible compared to the difference in value of the relevant options. Thus, the agent will not be tempted to win her bet by selecting the less valuable option. This strategy works as long as the agent is not indifferent between all options.[9] Second, we can design the betting mechanism so that the agent sets the odds and the bettor sets the stakes, without the agent knowing whether she will win or lose the bet if the alternative she bets on is chosen. The information available to the agent will be sufficient for measuring her subjective beliefs about her choice disposition, but because she does not know what choices will make her win or lose the bet, she has no incentive to adjust her choice disposition (see Rabinowicz [2002] for a discussion of this point).

The upshot is that Spohn and Levi are right that the agent's betting dispositions will sometimes affect the measurement process used for making accurate self-predictions, but the relevance of this observation should not be exaggerated. If we are aware of the problem, we can adjust the measurement process so that the problem they identify will not arise. The underlying phenomenon we wish to measure – the agent's subjective degree of belief that she makes a given choice – surely exists. So this is just a matter of performing the measurement in the correct way.

2.8 Wise Choice

Earlier we defended the manner in which the sophisticated conception of rational choice appeals to the *probabilities* of futures choices. This part of the theory is novel. In previous discussions of sophisticated choice, it has been assumed that the agent is able to predict *with certainty* his future choices, except perhaps where there are ties for maximal value. By allowing sophisticated agents to make probabilistic self-predictions, we make the theory applicable to a broader range of choice problems.

The sophisticated conception of sequential choice is often (e.g., McClennen 1990) defined as also satisfying a second feature: historical separability. This requires that the rational permissibility of a choice not depend on the history (e.g., what choices were made or what plans were adopted) prior to that choice node. As indicated in our discussion of the resolute conception, we agree with McClennen that this is an unreasonable requirement. Rational choice can be sensitive to what the past was like in virtue of having rational preferences that are historical (e.g., rationally preferring coffee after a meal but scotch before a meal). Although the sophisticated conception is not always understood to require satisfaction of the separability condition, we will here accept McClennen's definition and therefore reject the sophisticated conception so understood.

---

[9] See Peterson (2006) for a detailed discussion.

Fortunately, there is already a name for the conception of rational choice that (1) appeals to the agent's predictions about her future choices but (2) does not require satisfaction of historical separability. This is the *wise choice* conception, as introduced and discussed by Rabinowicz (1995, 1997, 2000, 2017). It allows, but does not require, that rational choice may depend on what the past was like and, in particular, on what plans were rationally adopted. It holds that rational choice in a given choice situation is a matter of choosing an option with best prospects in light of one's choice dispositions in the future. If one's rational preferences are nonhistorical, then this will be equivalent to the sophisticated conception. If one's rational preferences give lexical primacy to compliance with plans rationally adopted in the past, then, for perfectly rationally resolute agents, this will be equivalent to the resolute conception. Finally, if one's preferences give some finite weight to compliance with plans rationally adopted in the past, then the wise conception will, like the resolute conception, tend to favor compliance with such plans. It will not, however, always favor such compliance because it will be but one of several competing considerations for what is best.

Like the sophisticated conception, the wise conception has been characterized as involving backward induction on one's future choice (e.g., Rabinowicz 1995). Our discussion of sophisticated choice, however, makes clear that (1) backward induction is not always possible, but (2) this is not a problem because the core idea is captured by reasoning based on the probabilities of one's future choices. So the wise conception, we believe, should be so understood. Wise agents believe that there is no fundamental difference between probabilities ascribed to one's own future choices and probabilities ascribed to future events.

Rabinowicz (2017) suggests that there is a difference between wise choice and sophisticated choice with respect to how they *predict* future choices when the agent expects her future motivational preferences to become distorted relative to her current rational preferences (e.g., Ulysses' preferences as he sails close to the Sirens or the preferences of a person about to get drunk). Following Machina (1991), Rabinowicz holds that commitment to a historical separability condition (as in sophisticated choice) requires *predicting future* preferences on the basis of current unconditional preferences. McClennen's historical separability condition, however, does not require this. It does not address how to predict future choices. It merely asserts that the rational permissibility of a choice does not depend on what the past was like. Both the sophisticated and the wise choice conceptions, as we understand them, judge the *rational permissibility* of current and future choices relative to the rational preferences at the time of the assessment (e.g., current rational preferences). Moreover, both hold that *predictions* of future choices are based on the best available evidence about future motivational preferences. So, in our understanding, sophisticated choice and wise choice differ only with respect to whether a preference for complying with rationally adopted plans can be rational.

The upshot of all this is that the probabilistic account of wise choice that we propose seems able to embrace the best parts of the sophisticated analysis (sensitivity to how the agent is likely to behave in the future) and the best parts of the resolute analysis (sensitivity to what plans were rationally adopted in the past, to the extent the agent's motivational preferences are so sensitive).

2.9 Two Objections

A standard objection to the sophisticated account is that it can judge a sequence of choices to be rationally permissible even when dominated by some alternative feasible sequence of choices. This objection could be raised against the wise analysis too if the agent's rational preferences are historically insensitive. Moreover, the objection applies even if the agents have no *akrasia* (as we will assume for illustration).

Consider, for example, Figure 2.3, and assume that (1) where no choice has a sure outcome, the agent rationally prefers the choice with the highest expected monetary payoff, and (2) the agent prefers $1 million for sure to a 10/11 chance of $5 million and a 1/11 change of $0. Sophisticated agents (i.e., nonhistorical wise agents) are rationally required to go "Up" at the second choice node (for the certainty of $1 million). At the first node, however, the agent is rationally required (and hence permitted) to choose "Down" – because (1) she predicts with certainty that she will choose "Up" if she reaches the second choice node, (2) at the first choice, node neither "Up" nor "Down" has a certain prospect, and (3) "Down" has higher expected value than "UpUp." This violates a standard dominance condition because going "UpDown" is better for the agent than "Down," no matter what chance events (*E* or *F*) occur.

We see no reason to worry about this alleged violation of the dominance principle. For nondynamic choice (with just one choice situation), the standard dominance principle is sound. For dynamic choice, however, one must distinguish between different dominance principles, only some of which are sound. In its most general sense, (weak) dominance occurs when (1) for all possible events, one option has outcomes that are at least as good

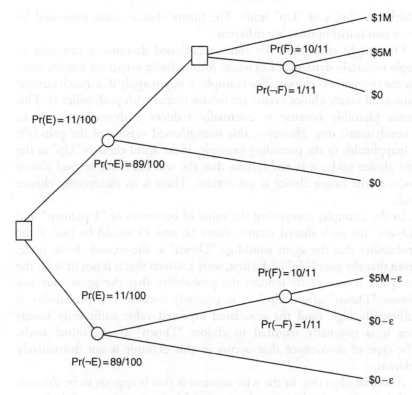

Figure 2.3 Alleged failure of dominance
(After Rabinowicz 1995, 2017.)

as those of a second option, and (2) for some possible events, it has better outcomes. In the nondynamic case, the only relevant events are the *chance events*, but this is not so in the dynamic case. One must also consider the *future choices* the agent might make. This gives rise to different dominance principles, only some of which are plausible. This violation of dominance, we claim, is only a violation of an implausible dominance principle. Let us explain.

A very weak dynamic dominance principle holds that if two options (choices or choice sequences) are defined on the *same set of chance and future choice events*, and one dominates the other relative to those events, then the latter is not a rationally permissible choice. This is extremely plausible. The preceding example, however, is not of this kind because the choice of "Down" initially does not lead to the second-choice node to which the choice of "Up" leads. The future choice events generated by these two initial options are different.

One might strengthen the above-mentioned dominance principle to apply in certain dynamic cases where future choice events are not the same for the two current choices. For example, it might apply if, for each current choice, all future choice events are *certain* (occur with probability 1). This seems plausible because it essentially reduces a dynamic choice to a nondynamic one. However, this strengthened version of the principle is inapplicable to the preceding example. If the agent chooses "Up" at the first choice node, it is not certain that she will reach the second choice node, so the future choice is not certain. There is an intervening chance node.

In the example, comparing the value of outcomes of "UpDown" and "Down" for each shared chance event (*E* and *F*) would be fine if the probability that the agent would go "Down" at the second choice node, given that she goes "Up" at the first, were 1. Given that it is not (it is 0), the comparison incorrectly ignores the probability that the agent may not choose "Down" after "Up." It is precisely because this probability is sufficiently high (and the associated expected value sufficiently lower) that it is rationally required to choose "Down" at the initial node. The type of dominance that occurs in this example is not normatively relevant.

A second objection to the wise account is that it appears to be *dynamically inconsistent,* in the sense that it can (1) judge a sequence of choices to be rationally *required* relative the first choice node of the sequence but (2) judge continuation of the sequence to be rationally *impermissible* relative to a later choice node of that sequence. For example, for any given decision tree, it is dynamically inconsistent to judge a plan "UpDown" to be a rationally required sequence relative to the starting choice node but to judge "Down" to be rationally impermissible relative to the second choice node reached after choosing "Up" at the first node (see McClennen 1990: sec. 7.2 and generally for more discussion).

Consider, for example, Figure 2.4. We assume here that a prospect is better, relative to the agent's rational preferences, just if it has higher expected monetary value. We further assume that the agent suffers from *akrasia* and has a disposition to take irrational risky gambles. Suppose that he predicts, at the first choice node, that if he goes "Up," he will go "Down" at the second node with probability 0.99 (because he is prone to take irrational gambles). The expected value of going up at the first node is therefore 0.01 × \$3 million + 0.99 × 1/2 × \$5 million = \$2.505 million,

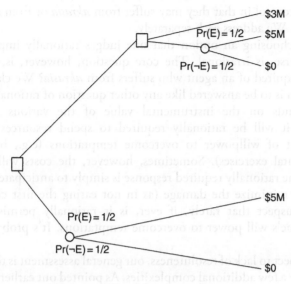

Figure 2.4 Alleged dynamic inconsistency

which is more than the expected value of going "Down" ($2.5 million). However, when the agent reaches the second node, the expected value of going "Up" is higher ($3 million is more than 1/2 × $5 million). Thus, despite his previous prediction to go "Down," the agent now decides to go "Up."

There is no dynamic inconsistency here. At the first node, the rational sequence of choices is "UpUp," and, at the second node, the rational choice is "Up." These sequences of choices are clearly consistent. There is of course an "inconsistency" between the *predicted* choice at the second node and the *rationally permissible* choice at the second node. This is, however, not a problem. This is precisely what we would expect for a wise agent predicting akratic behavior. An agent may predict akratic choice, and the rationality of his first choice would be based on that prediction, but it's still true that the rational choice at the second node is "Up." Stated otherwise, the wise agent predicts "Down" at the second node, but he is not endorsing the rationality of doing so. Hence there is no violation of dynamic consistency.

2.10 Self-Control for Wise Agents

Before concluding, we briefly comment on the value of self-control (willpower) in the broad sense) for wise agents. Wise agents need not have perfect self-control in that they may suffer from *akrasia* or from a lack of resoluteness. We address each separately.

Akrasia (choosing an option that one judges rationally impermissible), of course, is irrational. The core question, however, is, what is rationally required of an agent who suffers from *akrasia*? We claim that this question is to be answered like any other question of rational choice. It all depends on the instrumental value of the various options. Sometimes it will be rationally required to spend resources on the development of willpower to overcome temptations (e.g., by doing various mental exercises). Sometimes, however, the costs will be too great, and the rationally required response is simply to anticipate cases of *akrasia* and minimize the damage (as in not eating the first chocolate bar). We suspect that rarely, if ever, is it rationally permissible to *maximize* one's will power to overcome temptations. It's probably just too costly.

With respect to lack of resoluteness, our general assessment is the same, but there are a few additional complexities. As pointed out earlier, there is a distinction between *unconditional* resoluteness and *rational* (conditional) resoluteness. Unconditional resoluteness is a disposition to comply with adopted plans, no matter what. Rational resoluteness is a disposition to comply with plans that were rationally adopted and for which the rational permissibility of the adoption has not been undermined by new, unanticipated information. It's rarely rationally permissible to be perfectly *unconditionally* resolute because plans can be adopted irrationally and unanticipated information can undermine plans that were rationally adopted. Indeed, it's probably rarely rationally permissible to be *strongly* unconditionally resolute. The exceptions are bizarre cases where the world rewards such dispositions. Rational resoluteness, in contrast, is, to the extent that it is feasible, often rationally permissible. So we shall focus on it.

Rational resoluteness comes in varying degrees, which can be understood in terms of the probability that the agent will comply with non-undermined, rationally adopted plans in light of various benefits of noncompliance. *Perfect* rational resoluteness is the case where the agent is certain to so comply. In general, it is instrumentally valuable to develop one's rational resoluteness because this helps one to overcome (1) future *akrasia*, (2) future choices that are rational relative to one's future values but are irrational relative to one's present values because (a) of anticipated changes in basic values (e.g., with age) or (b) no changes in basic values, but basic values, at a time, are (i) *insensitive to the past* but (ii) *sensitive to the temporal order of future events*,[10] and (3) future choices that fail to adequately promote one's values because of a deterioration in the completeness or truth of one's beliefs (e.g., due to declining mental powers),

How, then, can one increase one's rational resolution?[11] One way is to increase one's capacity and disposition *not to reconsider* nonundermined, rationally adopted plans and thus allow them to be implemented. Second, one may wish to create, or strengthen, *a rational desire* to conformance with nonundermined, rationally adopted plans. To the extent that one's motivating preferences match one rational preferences, one will be more inclined to be rationally resolute. Third, some have argued (mainly McClennen 1990, 1997) that the mere act of "rational commitment" (an act of will) can make one rationally resolute or at least increase such resoluteness.[12] We are skeptical of this possibility, but we here leave this open. No doubt, there are other possibilities.

If some degree of rational resoluteness is feasible for an agent, it will often be instrumentally rational to cultivate such resoluteness. Of course, it does not follow that *perfect* rational resoluteness is feasible for real agents. Indeed, we are skeptical, but we leave this too open. Moreover, even if it is feasible, it may not be rationally permissible to develop (because of the costs involved).

In short, on the wise conception of rational choice, the degree of rational resoluteness that an agent is rationally required to have will be determined by instrumental considerations: the feasibility of developing it, the costs and benefits of doing so, and so on. Because we doubt that it will be feasible and instrumentally rational for any (current) human agent to become perfectly rationally resolute, we doubt that the resolute theory of rational choice applies to any (current) humans.

---

[10] For example, suppose that (1) one rationally prefers a future with eating chocolate followed by drinking coffee to one with the reverse order and prefers the latter to coffee at both times, but (2) if there is only one consumption event in the future, one rationally prefers eating chocolate to drinking coffee. One might now rationally decide to drink coffee first because one anticipates that one will later rationally decide to eat chocolate (since the past will be irrelevant at that point).

[11] See Holton (2009) and Bratman (2012) for insightful discussion of how resoluteness might be developed.

[12] Rational commitment is distinct from precommitment, which involves making some future choice infeasible (e.g., Ulysses tying himself to the mast) or altering the *external* costs of benefits of various choices (e.g., signing a contract to pay large sums of money if one smokes). Rational commitment leaves feasibility and external costs and benefits in place but is somehow supposed to be capable of modifying choice behavior.

2.11   Conclusion

We have proposed a theory of rationally permissible sequential choice that covers agents that have perfect self-control, agents who have partial self-control, and agents who have none. In this view, agents should ascribe probabilities to their own future choices (reflecting their dispositions for self-control) and at each point in time do whatever maximizes expected utility. This theory (1) agrees with the resolute conception of rational choice for agents who are perfectly rationally resolute in the evaluation-based sense (of having a lexically primary rational preference for compliance with nonundermined, rationally adopted plans), (2) agrees with the sophisticated conception of rational choice for agents who have nonhistorical motivational preferences (and hence no rational resoluteness), and (3) covers agents with imperfect rational resoluteness and/or *akrasia*.

References

Bratman, M. 2012. Time, rationality, and self-governance. *Philosophical Issues* 22:73–88.

Carlson, E. 2003. Dynamic inconsistency and performable plans. *Philosophical Studies* 113:181–200.

Hammond, P. J. 1988. Consequentialist foundations for expected utility. *Theory and Decision* 25:25–78.

Holton, R. 2009. *Willing, Wanting, Waiting*. Oxford: Oxford University Press.

Kavka, G. 1983. The toxin puzzle. *Analysis* 43:33–36.

Levi, I. 1989. Rationality, prediction, and autonomous choice. *Canadian Journal of Philosophy* 19:339–62 [reprinted in Isaac Levi, *The Covenant of Reason* (Cambridge: Cambridge University Press, 1997), pp. 19–39].

Machina, M. 1991. Dynamic consistency and non-expected utility. In *Foundations of Decision Theory: Issues and Advances*, ed. Michael Bacharach and Susan Hurley (pp. 39–91). Oxford: Blackwell.

Mele, A. R. 1992. *Irrationality: An Essay on Akrasia, Self-Deception, and Self-Control*. Oxford: Oxford University Press.

McClennen, E. F. 1990. *Rationality and Dynamic Choice*. Cambridge: Cambridge University Press.

McClennen, E. F. 1997. Pragmatic rationality and rules. *Philosophy and Public Affairs* 26:210–58.

Peterson, M. 2006. Indeterminate preferences. *Philosophical Studies* 130:297–320.

Rabinowicz, W. 1995. To have one's cake and eat it, too: sequential choices and expected-utility violations. *Journal of Philosophy* 92:586–620.

Rabinowicz, W. 1997. Wise choice: on dynamic decision-making without independence. In *Logic, Action, and Cognition*, ed. E. Ejerhed and S. Lindström (pp. 97–112). Dordrecht: Kluwer.

Rabinowicz, W. 2000. Money pump with foresight. In *Imperceptible Harms and Benefits*, ed. Michael J. Almeida (pp. 123–54). Dordrecht: Kluwer.

Rabinowicz, W. 2002. Does practical deliberation crowd out self-prediction? *Erkenntnis* 57:91–122.

Rabinowicz, W. 2017. Between sophistication and resolution – wise choice. In *Handbook of Practical Reason*, ed. Kurt Sylvan. London: Routledge.

Schick, F. 1986. Dutch books and money pumps. *Journal of Philosophy* 83:112–19.

Spohn, W. 1977. Where Luce and Krantz do really generalize Savage's decision model. *Erkenntnis* 11:113–34.

CHAPTER 3

Rational Plans

Paul Weirich[*]

Often self-control is rational and yet seems to require forgoing what one most wants. Do cases of rational self-control overturn decision theory's standard of utility maximization? I argue that they do not, investigating especially self-control in the execution of plans. My defense of utility maximization treats cases drawn from McClennen (1990), Rabinowicz (1995), Arntzenius, Elga, and Hawthorne (2004), and Elga (2010).

3.1 Rationality

Decision theorists adopt various accounts of rationality. For clarity, this section briefly presents the account I adopt. Weirich (2010: chap. 3) elaborates and supports it.

Rationality may be either a mental capacity or an evaluative benchmark for products of the mind – such as beliefs, desires, and choices. I treat rationality taken as an evaluative benchmark and consider whether choices meet the measure. Choices that fall short of the measure, and so are irrational, are blameworthy. Because blame is appropriate only when a means of escaping it exists, in every decision problem some choice is rational. Also, rationality is sensitive to an agent's abilities and circumstances. It demands more of an adult than it demands of a child and demands more of an adult able to reflect than of one hurried and distracted.

An agent may want a proposition to be true and may want one proposition more than another to be true. Desire that a proposition be true comes in various strengths. An agent's utility assignment to a proposition

* For helpful comments, I thank an anonymous referee and participants at the 2016 TAMU conference on self-control, especially Martin Peterson and Peter Vallentyne. I also received valuable comments on precursors of this chapter at the 2017 Pacific Division Meeting of the American Philosophical Association, especially from my commentator, Kino Zhao, and at the 2017 Formal Ethics Conference, especially from Wlodek Rabinowicz.

represents the agent's strength of desire that the proposition be true. It represents an agent's preferences and intensities of preferences if the agent is rational and ideal; such an agent wants a proposition to be true if and only if she prefers its being true to its being false. Although a proposition's utility for an agent is a subjective evaluation of the proposition's realization, if an agent's strengths of desires are rational, then the agent's utility assignment depends on the agent's basic goals and information, which settle the agent's nonbasic information-sensitive goals. Whether it is rational for an agent to follow her preferences depends on their status. If they are irrational, then a choice following preferences may be irrational. In the idealized cases I treat, agents have rational preferences. Utility maximization is equivalent to choosing according to rational preferences held, all things considered.

A decision problem gives an agent a set of options that are in the agent's direct control; the agent can perform them at will. In a decision problem with an option of stable maximum utility, rationality requires a cognitively ideal agent in ideal circumstances for deliberation to maximize utility, assuming that the agent's utility assignment is rational. Decision theory advances its standard of utility maximization for cases that are ideal, except possibly involving agents without complete information. When information is incomplete, an option's utility equals its expected utility (computed using a probability assignment that is relative to the agent's evidence) if the agent is cognitively ideal and rational. When an agent's information and goals do not single out a probability assignment and a utility assignment for possible outcomes but instead admit multiple pairs of a probability assignment and a utility assignment, a relaxation of the standard of utility maximization requires maximizing expected utility given some admissible pair.

3.2 Self-Control

This section presents this chapter's account of self-control. It does not attempt to resolve debates about the nature of self-control but just states the chapter's assumptions for clarity.

A common simplifying assumption takes choices to reveal preferences, all things considered. I dispense with this simplifying assumption to allow for phenomena such as weakness of will. An agent may act contrary to preferences held, all things considered. A desire that ignores some considerations may prompt an act. This can happen if countervailing considerations are insufficiently vivid or do not receive sufficient attention for preventing the act or if the agent does not appreciate the force with which they count against performing the act. Various factors, including weakness of will, may prevent the set of pros and cons from playing its role in settling the act's performance.

Self-control brings reflection to bear on action. It opposes impulsivity, weakness of will, and giving-in to temptation.[1] I count as self-control resisting a pure time preference that urges taking a good now rather than a greater good later just because the greater good is later. The attraction to the good now need not count as temptation, and being moved by it need not count as weakness of will. Still, because nearness makes a good vivid and thereby augments its motivational force, an agent exercises self-control when passing over a good now for the sake of a greater good later. Self-control may emerge from a reflective, circumspect state of mind.

An agent may use self-control to persist in a plan despite reasons to deviate from that plan. The reasons for deviation need not count as temptations, and deviation from the plan need not arise from weakness of will. Deviating from the plan may, for example, be vacillation rather than giving in to temptation, and it may arise from indecisiveness rather than from weakness of will.

An act that constitutes self-control typically influences a future choice. It may, for instance, be looking away from travel ads so as not to be tempted to take time off work. On compatibilist grounds, I assume that an agent may influence a choice without removing choice. Eating a potato chip may make me eat another, although I'm still free not to eat another. Deliberation treats as undetermined the options in a current decision problem but recognizes that current acts may determine future acts.

An agent exercises self-control by influencing his choices. His self-control may influence without settling a choice. For example, his taking pills may reduce an urge to smoke without ensuring that he will choose not to smoke. Ulysses had himself bound to the mast of his ship to control his acts. Binding is a form of self-control; it is an extreme case of influencing a choice. In general, binding removes an option, penalizes an option, or in some other way settles a choice. Ulysses removed his option of jumping into the sea to join the Sirens.

---

[1] Although distinguishing weakness of will and giving in to temptation are not crucial for the chapter's arguments about self-control, the phenomena differ. Suppose that I realize that I ought to rake the leaves but do not because I am lazy and insufficiently motivated to do the job. I exhibit weakness of will even if nothing, such as a desire to watch football, tempts me away from the job. Weakness of will without temptation is not giving in to temptation.

Some forms of binding ensure performing an act even if the act is irrational, whereas other forms of binding change the consequences of acts and so change the acts that maximize utility. Taking a pill that generates an act may bind one to the act's performance even if the act is irrational because it does not maximize utility. Entering a contract may bind one to an act's performance by adding penalties to not performing the act so that performing the act maximizes utility and is rational.

In cases of binding, the exercise of self-control generates another act with a distinct realization and distinct consequences. Suppose that I make myself stop eating potato chips by imposing a penalty on continuing. Imposing the penalty differs from my stopping. Rationality need not be transmitted from an act of self-control to its progeny.[2]

3.3 Rational Self-Control

An exercise of self-control may influence a choice by affecting deliberation and changing beliefs and desires without restricting or penalizing options. It may make deliberation circumspect so that an agent maximizes utility and thereby pursues goals that she adopts, all things considered. She is not sidetracked by the motivational force of desires arising from some, but not all, considerations. Utility maximization by itself manifests self-control in the resolution of a decision problem. In an ideal decision problem with an ideal agent, rational self-control by circumspection both maximizes and generates a maximizing act. Similarly, self-control through binding by contract creates an incentive to perform an act that may make the act maximizing. In many cases, both entering the contract and honoring the contract are maximizing.

Self-control may serve bad ends and may foreseeably have bad consequences. It may focus attention on an act's favorable features to encourage the act despite the act's being bad on balance. An agent may make himself stubbornly adhere to a plan although the reasons for the plan have vanished. This type of self-control is irrational. However, suppose that the reasons for a plan are constant and following the whole plan maximizes utility with respect to rival plans, although the plan requires steps that do not maximize utility. Then self-control may yield adherence to the plan when a step goes against all-things-considered preferences by focusing

---

[2] The act of making myself stop and the act of stopping have a common realization if the mechanism of self-control is just the act of stopping – I can make myself stop by stopping. Then, although making myself stop does not cause me to stop, making myself stop is still self-control, and making myself stop is still distinct from my stopping.

attention on considerations that support the plan. In this case, is self-control rational even though it generates an irrational act that flouts all-things-considered preferences? I look for an objection to utility maximization by considering this question and respond to it after first describing the sequences of choices that execution of plans produces.

3.4   Sequences of Choices

A decision tree, as in Figure 3.1, may represent a sequence of choices by the same agent. A decision node, which a square represents, stands for a decision problem, and the branches the node starts stand for options in the decision problem. The terminal node of a branch going from start to finish stands for the final outcome of the sequence of choices that the branch represents. The agent's preferences over final outcomes, assuming that these preferences exist, ground the agent's preferences over sequences of choices that the decision tree specifies.

A decision tree *adequately* represents decisions just in case it represents all considerations relevant to the decisions. A tree represents a consideration even if it does not mention the consideration, provided that it mentions another consideration that covers the first consideration given background assumptions about the agent, the decision problem, and the agent's circumstances. For example, a terminal node adequately represents the outcome of a bet as simply having made the bet, even if it does not mention the events that lead to winning the bet, provided that the tree's

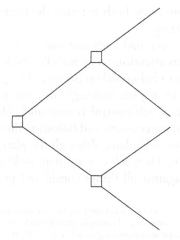

Figure 3.1 A decision tree for a sequence of two choices

representation of the bet's utility derives from the probabilities and the utilities of winning and of losing the bet.

Consider an adequate decision tree with only decision nodes; the agent completely controls his path through the tree. In the ideal cases I treat, each decision problem in the tree has an option of maximum utility, and the agent is ideal, knows where he is in the tree, and has perfect information about his history in the decision tree. At a decision node, a *sophisticated*, as opposed to a *myopic*, agent uses information about later choices to identify an option's consequences. In a decision tree for an agent, I assume that the agent is sophisticated and, moreover, foresees his choices throughout the tree. A sophisticated agent may predict his choices using knowledge of his preferences and his method of using his preferences to select an option. He knows the final outcome that follows from his choice at a node because he knows his choices at future nodes. The agent's foreknowledge of his choices follows from his being cognitively ideal if he foresees the grounds of his choices. By assumption, he anticipates the information he will have at a node if he reaches the node, and because his tree has no chance nodes and he foresees his choices, he acquires no new relevant information at the node.

Preferences obtain at times and may compare either options or sequences of options. At each node, consistency requires that preferences among options agree with preferences among sequences of options in present and future decision problems. Suppose that an agent during progress through his decision tree has the same basic goals and the same information concerning the outcomes of his choices at each decision node. Then rationality permits the agent to have, and this section assumes that he has, a constant preference ranking of sequences of choices in the decision tree. His ranking of options at a node agrees with his constant ranking of sequences of choices. If rational, his act at each node is part of the best sequence available according to the agent's preference ranking of sequences.

By following preferences at each node, an agent realizes the best sequence of choices and, if several sequences are best, realizes a selected one of the best, which I call the best to simplify terminology. At each node, the agent makes a choice that belongs to the best continuation of the sequence started. Given the agent's confidence that he will behave this way throughout the tree, maximizing utility at each node produces the best sequence.

Backward induction establishes this result. At the last node, participating in the best continuation of the sequence realized up to that node maximizes; at the second to last node, participating in the best continuation maximizes; and so on up to and including the first node. At the first node, participating in the best continuation entails participating in the best sequence. Hence, at each node, participating in the best continuation entails participating in the best sequence. At a node along the path that represents the best sequence, participating in the best sequence maximizes utility among options at the node given the agent's knowledge of his other choices in the tree.

This argument that stepwise maximization yields a maximizing sequence fails if some sequences are not compared, if some branches are infinitely long, or if the decision tree includes chance nodes or decision nodes for other agents. The following sections examine cases with these features and consider whether, when maximizing steps fail to generate a maximizing sequence, rational self-control challenges the standard of utility maximization. Because a challenge arises only if agents and their decision problems meet the standard's idealizations, the cases treated preserve these idealizations. In these cases, the agent is cognitively ideal and complies with rationality's requirements, except possibly the requirement to maximize utility at a decision node.

Using a utility-maximizing sequence of choices with a nonmaximizing choice at some decision node to argue against the standard of utility maximization at the decision node assumes that the agent has a constant (although possibly incomplete) preference ranking of final outcomes of sequences of choices. It also assumes that arrival at a decision node brings no unanticipated information so that the preference ranking of options at a decision node is constant (although possibly incomplete).

The constancy of preferences among options at a decision node presumes that an agent may have a preference between options not currently available. An agent prefers one option to another in a decision problem not currently faced if, when the agent supposes confronting the decision problem, he prefers the first option to the second. His current preference between the two options depends on his preference in a hypothetical decision problem. For a rational ideal agent, this preference is equivalent to (but not defined as) a preference conditional on confronting the decision problem. For the conditional preference, the agent indicatively supposes, as an assumption of the agent's reasoning, confronting the decision problem. I assume that an agent at the start of a sequence of decision problems, for each decision problem, has in the sense explained a preference ranking of the options in the decision problem. Because the ranking is constant throughout the sequence of decision problems, initial preferences among the options in a decision problem settle preferences conditional on facing the decision problem, and these conditional preferences become nonconditional preferences when the decision problem arises.

3.5 Resolute Choice

In the sequential version of the prisoner's dilemma in Figure 3.2, the first agent decides whether to help the second agent, and then the second decides whether to help the first. The pair of numbers at a terminal node lists the final outcome's utility for the first agent and, next, the final outcome's utility for the second agent. The first agent helps the second only if convinced that the second will reciprocate. The second openly commits herself to helping the first and thereby wins the first's help. However, when the time comes for the second to help the first, the second's helping does not maximize utility for the second. Maximization urges her to break her commitment. Her resolutely honoring her commitment brings her help but requires an act that does not maximize utility. McClennen (1990) holds that the second agent's helping, because it secures the first agent's helping, is rational despite not

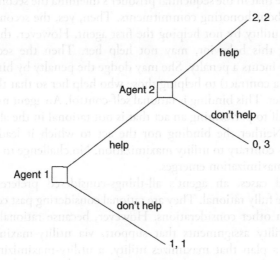

Figure 3.2 A sequential prisoner's dilemma

maximizing utility. He maintains that an agent who resolutely follows rational plans sometimes acts contrary to utility maximization.

The sequential prisoner's dilemma puts aside the argument of Section 3.4 that maximizing steps yield a maximizing sequence by introducing a decision tree for an agent with nodes for decisions by other agents. The game tree it creates differs from the decision trees that Section 3.4 treats. Information about an agent's choice may influence another agent's choice.

Although I assume that an agent's preferences are constant and independent of past choices, the past affects the characterization of future events. It affects the nature of events that form the consequences of a future choice. Thus the consequences of a future choice may depend on a past choice. For example, because of a past commitment, some future choice may constitute honoring a commitment.

The traditional standard of utility maximization, as advanced by Daniel Bernoulli, assumes that an option's possible outcomes are comprehensive and include everything that matters to the agent. Suppose that the sequential prisoner's dilemma excludes honoring a commitment from the possible outcomes of choices. If possible outcomes include this consequence, then the second agent's helping, because it honors a commitment, may maximize utility. The sequential prisoner's dilemma, using comprehensive utility assignments to options, does not show that rational self-control fails to maximize utility without further assumptions.

Suppose that in the sequential prisoner's dilemma the second agent does not care about honoring commitments. Then, yes, the second agent may maximize utility by not helping the first agent. However, the first agent, predicting this behavior, may not help her. Then the second agent's rationality incurs a penalty. She may dodge the penalty by binding herself (say, with a contract) to helping those who help her so that the first agent will help her. This binding is rational self-control. An agent may rationally bind herself to performing an act that is not rational in the absence of the binding. Neither the binding nor the act to which it leads, given the binding, is contrary to utility maximization. No challenge to the standard of utility maximization emerges.

In ideal cases, an agent's all-things-considered preferences among options are fully rational. They are rational considering past commitments along with other considerations. However, because rationality does not require utility assignments that support, via utility-maximizing steps, following a plan that maximizes utility, a utility-maximizing plan may have a step that does not maximize utility. Is it rational for an agent to choose contrary to preferences because of a past commitment to a maximizing plan of action? McClennen (1990), as I interpret him, claims that resoluteness may yield choices contrary to preferences and so contrary to utility maximization. A resolute agent's choices do not follow preferences at each step in a plan's execution but instead follow the plan at the top of the agent's preference ranking of plans at the start of the plan's execution.[3]

Support for self-control that does not maximize utility uses (1) utility maximization as a standard of rationality for plans and (2) belonging to a rational plan as a sufficient condition for an act's rationality. Belonging to a rational plan is not sufficient for an act's rationality if the act is an incidental part of the plan or if the plan's adoption has good consequences that do not follow from the plan's execution. However, my main objection targets the assumption that a rational plan maximizes utility. I examine the rationality of executing a plan and, more precisely, the rationality of the sequence of choices that constitutes execution of the plan. I treat plans that are rational because their content is a rational sequence of choices and ask what makes a sequence of choices rational.

Using utility maximization to evaluate a sequence of choices requires specifying the set of sequences compared, usually sequences of choices during the same period. A decision tree for an agent may specify the comparison set. Evaluating choices by evaluating the sequences of choices to which they belong generates inconsistency in evaluations of choices because a choice may belong to many sequences, some rational and others not rational. Suppose that a single act counts as a sequence, that a sequence of acts counts as an act, and that a sequence is rational if and only if it maximizes among rival sequences. Then an act is rational if and only if maximizing, but also if and only if it is part of a larger maximizing sequence. Inconsistency arises when a maximizing act is part of a nonmaximizing sequence or when a nonmaximizing act is part of a maximizing sequence.

Justification of the standard of utility maximization restricts the standard's application to a decision problem's resolution. Its justification assumes a decision problem in which an agent has direct control over each option and can realize it at will. An agent does not have direct control over a sequence of choices because she cannot realize the

---

[3] An alternative definition of a resolute choice requires that the agent follow a plan that is Pareto optimal for time slices of the agent. A definition that does not treat as agents time slices of the agent strengthens the case for a resolute choice.

sequence at will (she may die before completing the sequence). She has to wait for arrival of the time of a choice in the sequence before she can make the choice. Rationality does not evaluate a sequence by applying the standard of utility maximization but rather by evaluating the sequence's elements. A sequence of rational choices generates a rational sequence even if the sequence does not maximize utility. The rationality of choices makes a sequence rational rather than the rationality of the sequence making the choices rational.

Rationality guides action and targets first acts and then sequences of acts because an agent performs a sequence of acts by performing the acts in the sequence. Principles of rationality for acts have priority over principles of rationality for sequences of acts. To be action guiding, rationality must target first acts in a current decision problem; its options are the ones the agent confronts. Also, to be action guiding, rationality must be consistent in its evaluations of sequences and their elements. It cannot prohibit performing a sequence but permit performing each element of the sequence. For consistency, rationality's requirements cannot say that a sequence is irrational even though it contains only rational choices. If the sequence is irrational, then rationality requires changing it. The sequence may change only if an element changes. Hence, if all the elements are rational, rationality does not require changing any, and the sequence is rational.

Because a plan with rational steps is rational even if the plan fails to maximize utility, utility maximization does not make a plan rational, and belonging to a rational plan does not make an act rational. Rationality requires an agent to maximize utility at every stage in a plan's execution. Granting the usual idealizations about the agent, the agent's decision problem, and the agent's circumstances, rational self-control at a stage therefore maximizes utility at the stage, both in the exercise of self-control and in the act to which self-control leads.

In a sequence of decision problems, if an agent follows preferences in each decision problem, then the agent's sequence of choices is rational. If an agent has rationally adopted a plan and rationally exercises self-control to follow the plan, then, I assume, if the self-control is rational, it produces rational choices, and as this section argues, the rational choices make the sequence of choices rational. The sequence may be rational despite not maximizing among rival sequences.

Realizing a sequence of maximum utility is a multichronic requirement on choices, and rationality does not impose this requirement independently of its synchronic requirements on choices. Section 3.6 considers whether consistency among the choices in a sequence imposes on choices a different independent multichronic requirement.

3.6 Intertemporal Consistency

The rationality of a sequence of choices depends on its meeting rationality's consistency constraints. An argument that rational self-control may conflict with the standard of utility maximization claims that maintaining the consistency of a sequence of choices may conflict with, and takes precedence over, maximizing utility at a step in the sequence. Is some consistency constraint on the choices in a sequence independent of the standard of utility maximization?

Figure 3.3 presents a sequential version of Allais' paradox. The circles depict chance nodes, and the branches they start represent epistemic possibilities. Each possibility comes with a probability.

The figure depicts the agent's choice at a decision node by thickening the line representing the option chosen. At the start of the tree, the agent selects a chance for offer A, which comes if event E occurs. If the offer arrives, he takes the sure $3,000. The decision tree's inclusion of chance nodes violates a condition of Section 3.4's argument that maximizing steps

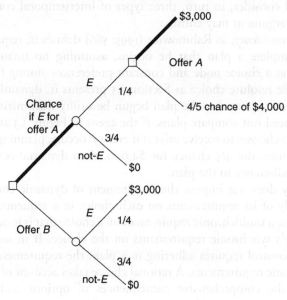

Figure 3.3 A sequential version of Allais' paradox

yield a maximizing sequence, but the case meets the argument's other conditions, including that the agent is rational, except possibly in a choice at a decision node.

The agent's choices considered in isolation from each other follow reasonable preferences at each node, but the preferences are inconsistent taken together, assuming that the agent cares only about money and ways of obtaining it. At the beginning, selection of the chance for offer A provides a $(1/4 \times 4/5)$ or $1/5$ chance of $4,000, which the agent prefers to a $1/4$ chance for $3,000. If the offer A arrives, he prefers $3,000 to a $4/5$ chance of $4,000. Using amounts of money as final outcomes, the first choice expresses the inequality $(1/5)U(\$4,000) > (1/4)U(\$3,000)$, so $(4/5)$ $U(\$4,000) > U(\$3,000)$. This is contrary to the inequality $U(\$3,000) > (4/5)U(\$4,000)$ that the second choice expresses.

Inconsistency makes an agent vulnerable to sure losses. Circumstances may make an inconsistent sequence of choices a nonmaximizing sequence. A slight change in offer B, described later, makes Figure 3.3's inconsistent sequence of choices a nonmaximizing sequence. Although it is possible to convert a requirement of intertemporal consistency among choices in a sequence into a requirement of utility maximization for sequences, an argument for using self-control to depart from preferences at a node may invoke a requirement of intertemporal consistency without making the conversion. I consider, in turn, three types of intertemporal consistency that such an argument may enlist.

Dynamic consistency, as Rabinowicz (1995: 586) defines it, requires that an agent complete a plan that he begins, assuming no unanticipated information at a choice node and constant preferences among final outcomes. Unlike resolute choice as Section 3.5 presents it, dynamic consistency does not require that the plan begun be utility maximizing, so its application need not compare plans. If the agent in Figure 3.3 at the first decision node chooses to receive offer A if event E occurs, planning to take, if offer A comes, the $4/5$ chance for $4,000, then dynamic consistency requires his adherence to the plan.

Rationality does not impose the requirement of dynamic consistency independently of its requirements on each choice in a sequence. A plan does not create a multichronic requirement on a choice that is independent of rationality's synchronic requirements on the choice. If in some cases rational self-control requires adhering to a plan, the requirement derives from synchronic requirements. A rational choice takes account of plans by considering the comprehensive consequences of options and not by supplementing synchronic requirements with the requirement of dynamic consistency.

For example, suppose that in Figure 3.3 offer *B* yields $3,001 if event *E* occurs, but the agent's choices remain the same because the increase from $3,000 to $3,001 is so small. The chance for offer *A* if event *E* occurs, and then, if offer *A* comes, the sure $3,000, together amount to $3,000 if *E* occurs. Therefore, the agent's sequence of choices does not maximize utility. The standard of utility maximization for each of the agent's choices by itself corrects this problem. Given the usual idealizations, the agent foresees his choices. Assuming that the agent foresees taking the sure $3,000 at the second decision node, the utility of taking the chance of offer *A* at the first decision node is less than the utility of taking offer *B*. A utility-maximizing agent therefore chooses offer *B* rather than the chance for offer *A*. Utility maximization at each step of a sequence of choices generates dynamic consistency. Rationality does not impose this type of intertemporal consistency as an independent requirement.

Conditionalization of preferences in a sequence of choices, as Rabinowicz (1995: 607–8) defines it, requires updating preferences using prior preferences among options in a decision problem, conditional on facing the decision problem, as nonconditional preferences among the options when facing the decision problem. An agent's sequence of choices is consistent in a sense just in case the choices follow preferences obtained by conditionalization. In Figure 3.3, the agent's preferences follow conditionalization if at the first decision node the agent's preference between $3,000 and a 4/5 chance of $4,000 given the second decision node is the same as the agent's preference between these options on arrival at the second decision node.

The case for conditionalization for preferences depends on the interpretation of a conditional preference. Suppose that the definition of conditionalization takes an agent's preferring a choice to a rival at a decision node, given that the agent is at the node, to rest on information the agent expects to have at the decision node, as in Section 3.4. Given this interpretation of a conditional preference, in Figure 3.3 the agent's preferences at the second decision node may, in fact, follow updating that uses conditional preferences. The agent may at the first decision node have a preference for the chance of offer *A* and also a preference, given that he is at the second node, for the sure $3,000. After arriving at the second decision node, updating then yields a preference for the sure $3,000. In this case, conditionalization does not block the choices Figure 3.3 depicts.

Suppose instead that the definition of conditionalization takes an agent's preferring an option to a rival at a future decision node to be the agent's currently preferring that he select the option rather than the rival at the future node. Then, in Figure 3.3, at the first decision node, a conditional preference for a 4/5 chance of $4,000 over a sure $3,000 at the second decision node amounts to the agent's at the first decision node preferring that he select that chance rather than the sure $3,000 at the second decision node. Assuming this conditional preference, conditionalization requires the agent nonconditionally to prefer at the second decision node the 4/5 chance of $4,000. It prevents the choice of $3,000.

Under this interpretation of a conditional preference, the case for making the conditional preference at the first decision node nonconditional at the second decision node is just the case for realizing the sequence of choices that the agent favors at the first decision node. No general requirement of rationality, applicable to agents whatever their basic goals, requires choosing to realize this sequence.

A third type of consistency among choices in a sequence, which I call *coherence*, requires that the choices all be as if maximizing expected utility according to a single utility assignment to final outcomes. As shown earlier, no utility function over final monetary outcomes makes the choices in Figure 3.3 maximize expected utility at each node, assuming that choices express preferences. However, rationality does not impose on a sequence of choices the requirement of intertemporal coherence. An agent's sequence of choices need not be as if maximizing expected utility according to a constant utility assignment to final outcomes. An agent may instead follow preferences among options at each decision node. Rationality imposes no multichronic requirement on choices that is independent of its action-guiding synchronic requirements on choices. All restrictions on sequences of choices arise from the synchronic requirements that guide an agent's choice at a time.

The sequence of choices in Figure 3.3 may seem irrational because the choices express a change in preference not justified by any change in circumstances. In particular, the choices express a change in the agent's appraisal of sequences of choices despite information at each decision node being anticipated and the utilities of final outcomes being constant. Although I accept the rationality of the sequence of choices in Figure 3.3, I note that a slight change in the case justifies a change in the agent's appraisal of final outcomes.

Suppose that the agent, instead of being neutral toward risk, has an aversion to risk so that final outcomes include risks. Despite the enrichment of final outcomes, the agent's choices still indicate a change of mind. At the first decision node, the agent passes over a 1/4 chance for $3,000 to have a chance to move to the second decision node, but at the second decision node his choice makes it the case that at the first decision node he has in effect selected a 1/4 chance for $3,000. With neutrality toward risk, the change of mind exercises a permission. With aversion to risk, it has a justification.

A justification of the change of mind comes from the agent's change in information. At the first decision node, the agent does not know that if he takes the chance for offer *A*, he will reach the second decision node. His movement from the first to the second decision node brings new information, even if it is information that the agent anticipates having should he reach the second decision node. Suppose that at the first decision node he prefers at the second decision node to take the 4/5 chance of $4,000. At the first decision node, he knows the information he will have at the second decision node if he reaches that node but does not know whether event *E* occurs. At the second decision node, the agent has the new information that *E* occurred. This new information may justify changing his preference between the option of $3,000 and the option of a 4/5 chance of $4,000. The preferences among options that an agent has at a decision node may differ from the preferences that he had earlier if at the node the agent has new, relevant information. At the second decision node, the agent's new information about *E* may support new preferences among options at the second decision node.

Conative attitudes such as desires are relative to a time. A rational agent may care about a pain to come but not about a pain past. Similarly, a rational agent may care about a risk to come but not about a risk past. Consider a risk grounded in the probability that the agent assigns to a bad event because of evidence concerning the bad event. The agent may lose his aversion to the risk after it passes; the aversion arises from uncertainty, and the uncertainty's resolution removes the risk and the aversion. The movement of a risk into the past may justify a change in preferences as an agent progresses through a sequence of choices. In Figure 3.3, at the first decision node, the agent's aiming for $4,000 involves a risk. The risk undertaken is part of the final outcome of each option of the second decision node. Passing through the chance node for the event *E* brings information about *E* that affects the agent's attitude to that risk, the agent's ranking of final outcomes, the agent's ranking of sequences of choices, and the agent's preferences among the options at the second decision node. Moving past the chance node and learning whether event *E* occurs may provide a reason for reversing an earlier conditional preference between $3,000 and a 4/5 chance of $4,000. Successfully passing the risk of not receiving offer *A* may justify a change in preferences among final outcomes and the sequences of choices that produce final outcomes. As the agent moves along a branch from the first decision node through the second decision node to the branch's final outcome, the final outcome does not change, and the agent does not acquire unanticipated information, but the agent's temporal position in the final outcome, a possible world, changes, and his evaluation of the final outcome changes because his new temporal position brings new relevant information.

In some cases, utility-maximizing choices may produce a sequence of choices that are intertemporally inconsistent in a sense. However, rationality imposes no requirement of intertemporal consistency that is independent of the requirement that each choice in the sequence maximize utility. This section concludes that rationality's consistency requirements for sequences of choices do not conflict with the standard of utility maximization.

3.7 Arbitrage

Section 3.4's argument that an agent's maximizing choices yield a maximizing sequence of choices assumes that in each decision problem the agent ranks all options. If probabilities of possible outcomes are imprecise, an agent may fail to rank all options. The standard of utility maximization generalized for such cases requires maximizing expected utility given some admissible pair of a probability assignment and a utility assignment. If choices meeting this permissive standard fail to yield a maximizing sequence, does rationality independently require self-control to create a maximizing sequence?

In an isolated decision problem, the permissive standard has strong support because, assuming that consequences are comprehensive, it responds to every relevant consideration. No consideration motivates demanding more than does the standard. To buttress this support, I address some objections to the permissive standard's application to isolated decision problems before considering its application within sequences of choices.

An option's utility settles the scope of its evaluation of the option. An option's evaluation is generally more precise if the scope of evaluation is narrow because this reduces uncertainty about the option's possible outcomes. By omitting some features of an option's outcome, the evaluation may gain precision because it need only assess the remaining features.

The separability (or utility independence) of the remaining features from those omitted justifies the omission.

The separability of future events from past events in evaluation of options makes the preference ranking of futures independent of the past. It grounds evaluating options by evaluating their futures, putting aside their common past, as Weirich (2015) explains. Uncertainty about the past makes imprecise an evaluation of an option's complete outcome, namely its world. Evaluation of the option's future, although also uncertain, is more precise in some cases because it considers fewer events. Evaluation of options becomes more precise by focusing on the future in a way that the separability of future events justifies.[4]

In some cases, narrowing the scope of an option's evaluation by increasing the precision of utility assignments narrows the set of permissible decisions. The permissive standard is inconsistent if it permits an option when the scope of evaluation is broad and evaluations are imprecise but prohibits the option when the scope of evaluation is narrow and evaluations are precise. Taking account of variation in the scope of evaluation of options, the permissive standard holds that a rational option maximizes with respect to some admissible pair of a probability assignment and a utility assignment for each admissible way of setting the evaluative scope of the utility function. Increasing the scope of evaluation of options does not, through increased imprecision in evaluations, create greater permission. If any admissible narrowing of the evaluation's scope makes an option nonmaximizing according to any admissible pair of a probability assignment and a utility assignment, then even if evaluations of wider scope make the option maximizing according to some admissible pair of a probability assignment and a utility assignment, the standard prohibits the option.

An objection to the permissive standard complains that the standard in some cases unjustifiably condones all choices simply because evaluations of options are imprecise. Suppose that in an agent's decision problem, for each option it is epistemically possible that the option's consequences are good and that its rivals' consequences are bad. In this case, each option maximizes utility using a utility assignment that gives the option high utility and gives its rivals low utility. Hence the permissive standard too easily allows each option.

---

[4] For simplicity, I assume that one world and one future with the option's realization are closest to the actual world and the actual future, respectively. The nearest world is the option's world, and the nearest future is the option's future.

The separability of basic goals places constraints on permissions despite uncertainty of an option's outcome. Imagine a choice whether to have now some pleasure that will not affect the future in ways one cares about. Whatever the past or future, the outcome is better if one chooses the pleasure now. The pleasure's separability from the past and future settles the choice. For every admissible scope of a utility assignment, choosing the pleasure maximizes according to some admissible pair of a probability assignment and a utility assignment.

Next, consider an agent making a sequence of two choices. Each choice is between two options that the agent does not compare despite reflection. The agent first selects option B given a choice between options $A+$ and B and then selects option A given a choice between options A and B. The option $A+$ is slightly better than the option A, so the pair of choices, which the permissive standard seems to allow, yields a loss of the difference between $A+$ and A. In fact, the permissive standard's idealizations, which include the agent's foresight, prevent this loss. An agent applying the permissive standard does not choose B over $A+$ if she foresees that she will subsequently choose A over B. Choosing B over $A+$, she knows, has a loss as a consequence. B has lower expected utility than $A+$ on every admissible pair of a probability assignment and a utility assignment.

Elga (2010) argues that when probabilities are imprecise, a sequence of choices, each of which maximizes with respect to an admissible pair of a probability assignment and a utility assignment, may not similarly maximize. Suppose that Sally is offered in sequence two gambles, as in Figure 3.4. Gamble A pays \$15 if H is false and costs \$10 otherwise. Gamble B pays \$15 if H is true and costs \$10 otherwise. By accepting both gambles, Sally guarantees a gain of \$5. However, if Sally assigns a probability interval [0.1, 0.8] to H and cares only about money (and ways of pursuing money), then the permissive standard allows her to reject each gamble even though rejecting both is irrational. Elga uses this case to argue against the permissive standard of rational choice.

Section 3.5 argues against a requirement of utility maximization for a sequence of choices that is independent of requirements on the choices in the sequence. Rationality's multichronic requirements emerge from its synchronic requirements. In Elga's case, no conflict between maximization in a sequence and maximization in its steps arises if Sally is an ideal agent who predicts her own choices. If she predicts her rejecting the second gamble, then she does not rationally reject the first gamble; the only reason for rejecting it is preparation to accept the second gamble and thereby stake chances for money on H's being true.

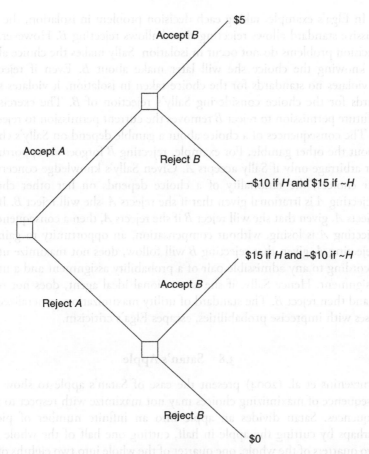

$5

Accept *B*

Accept *A*

Reject *B*

−$10 if *H* and $15 if ~*H*

$15 if *H* and −$10 if ~*H*

Accept *B*

Reject *A*

Reject *B*

$0

Figure 3.4 A sequence of offers

Sally's rejecting *A* and then rejecting *B* squanders an opportunity to gain money. Rejecting *A* is permissible if she foresees accepting *B*. Although after rejecting A, rejecting *B* is permissible because −$10 is worse than −$5, if Sally were to exercise this permission, given her foresight, she would have been mistaken to reject *A*. Thus, if she rejects *A* and then rejects *B*, either rejecting *B* was a mistake because it foregoes the opportunity to gain from gamble *B* or if it is not a mistake, then rejecting *A* was a mistake because it squanders the opportunity for arbitrage. The permissive standard, designed for ideal agents, survives Elga's objection that it leads to non-maximizing sequences of choices.

In Elga's example, taking each decision problem in isolation, the permissive standard allows rejecting A and allows rejecting B. However, the decision problems do not occur in isolation. Sally makes the choice about A knowing the choice she will later make about B. Even if rejecting A violates no standards for the choice taken in isolation, it violates standards for the choice considering Sally's rejection of B. The exercise of a future permission to reject B removes the current permission to reject A.

The consequences of a choice about a gamble depend on Sally's choice about the other gamble. For example, rejecting B forgoes an opportunity for arbitrage only if Sally accepts A. Given Sally's knowledge concerning her choices, the rationality of a choice depends on her other choice. Rejecting A is irrational given that if she rejects A, she will reject B. If she rejects A, given that she will reject B if she rejects A, then a consequence of rejecting A is losing, without compensation, an opportunity to gain \$5. Rejecting A, given that rejecting B will follow, does not maximize utility according to any admissible pair of a probability assignment and a utility assignment. Hence Sally, if she is a rational ideal agent, does not reject A and then reject B. The standard of utility maximization, generalized for cases with imprecise probabilities, escapes Elga's criticism.

3.8   Satan's Apple

Arntzenius et al. (2004) present the case of Satan's apple to show that a sequence of maximizing choices may not maximize with respect to rival sequences. Satan divides an apple into an infinite number of pieces, perhaps by cutting the apple in half, cutting one half of the whole into two quarters of the whole, one quarter of the whole into two eighths of the whole, and so on so that the pieces available as fractions of the whole are $1/2, 1/4, 1/8, \ldots$. Satan offers Eve the pieces one by one, and Eve decides, for each piece as it is offered, whether to take the piece or not, as in Figure 3.5. She benefits from each piece, but if she takes an infinite number of pieces, she is expelled from the Garden of Eden, a punishment that outweighs the benefit of the whole apple. In each decision problem, taking has more utility than not taking. Eve does better taking than not taking because one more piece does not move her collection from a finite to an infinite number of pieces. Nonetheless, taking all the pieces does not maximize utility because of the penalty it incurs.

Because the case involves an infinite sequence of choices, it does not meet the assumptions of Section 3.4's argument that a maximizing sequence of choices arises from choices that maximize. In an infinite

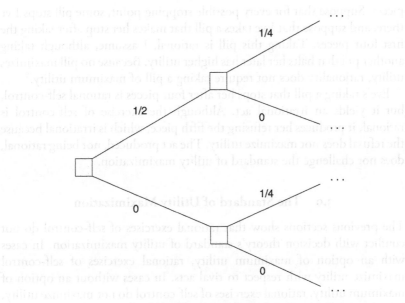

Figure 3.5 Satan's apple

sequence of takings, each choice maximizes utility, but the sequence itself does not maximize utility. Arntzenius et al. reason that each choice is rational because it is utility maximizing and that the whole sequence is rational because it has rational steps. Because no global standard of rationality for a sequence of choices exists independently of the standards for each choice, if each choice meets standards, then so does the sequence. Independent standards for sequences of choices, if there were such, would fail to motivate Eve's choices because she faces just one choice at a time, and only her preferences at a choice's time motivate her choice.

Arntzenius et al. use Satan's apple to make a point about binding, a form of self-control. They hold that Eve may escape the penalty for greed by binding herself to not accepting pieces after taking some finite number of pieces. She needs a means of binding herself or else her rationality in her choices brings her ruin. They do not describe a mechanism of binding. Some forms of binding make taking a piece after a selected stopping point incur a penalty so that taking the piece is not maximizing. Then rationality does not require taking the piece. However, Satan may block these forms of binding and leave available only binding in the form of taking a pill that makes Eve irrationally refuse each piece after taking some finite number of pieces. Suppose that for every possible stopping point, some pill stops Eve there, and suppose that Eve takes a pill that makes her stop after taking the first four pieces. Taking this pill is rational, I assume, although taking another pill that halts her later has higher utility. Because no pill maximizes utility, rationality does not require taking a pill of maximum utility.[5]

Eve's taking a pill that stops her after four pieces is rational self-control, but it yields an irrational act. Although the exercise of self-control is rational, it produces her refusing the fifth piece, which is irrational because the refusal does not maximize utility. The act produced, not being rational, does not challenge the standard of utility maximization.

3.9 The Standard of Utility Maximization

The previous sections show that rational exercises of self-control do not conflict with decision theory's standard of utility maximization. In cases with an option of maximum utility, rational exercises of self-control maximize utility with respect to rival acts. In cases without an option of maximum utility, rational exercises of self-control do not maximize utility, but the standard of utility maximization does not govern such cases. Therefore, these cases do not refute the standard.

In a decision problem with an option of maximum utility, when an agent makes a rational choice, the agent follows her preferences. Because she has a reason to follow her preferences, she has a reason to make the choice. Rationality requires following preferences rationally held all things considered, and utilities represent such preferences, so rationality requires utility maximization.

The standard of utility maximization governs a decision problem without restrictions on the nature of the options in the decision problem. Although the standard assumes idealizations about agents and their decision problems, the idealizations about decision problems accommodate options of any type; the options may include exercises of self-control.

---

[5] Theorists who assert the possibility of dilemmas of rationality may resist the last point. They may conclude that in this decision problem, every option is irrational. If in an agent's decision problem every option is irrational, then the agent ought not adopt any option. However, granting that the options in the decision problem are exhaustive, the agent must adopt one. Rationality is not action guiding if it declares that every option is irrational; its prohibitions are then impossible to observe. Section 3.1 takes rationality to be action guiding and so in compliance with the principles that "ought" implies "can" and that "ought not" implies "can not." Weirich (2010: 39, 100) argues against dilemmas of rationality.

This chapter does not advance decision principles for problems in which no option maximizes utility. Decision theory addresses such problems, but their subtleties are outside this chapter's scope.

The grounds for utility maximization apply to decision problems concerning acts of self-control. Nothing about such acts undermines the standard's support when applied to these acts. Although acts of self-control may have distinctive consequences, general methods of comparing options still apply. Rational acts of self-control conform to, and do not conflict with, the standard of utility maximization.

Rational exercises of self-control, such as binding, may, however, generate other acts that do not maximize utility even when some rival maximizes utility. In this way, rational self-control may lead to violations of the standard of utility maximization. Such violations, nonetheless, do not discredit the standard because the nonmaximizing act, to which rational self-control leads, is irrational. Utility maximization survives as a standard of rationality.

References

Arntzenius, F., A. Elga, and J. Hawthorne. 2004. Bayesianism, infinite decisions, and binding. *Mind* 113:251–83.

Elga, A. 2010. Subjective probabilities should be sharp. *Philosophers' Imprint* 10 (5):1–11. Available at www.philosophersimprint.org/010005/.

McClennen, E. 1990. *Rationality and Dynamic Choice: Foundational Explorations*. Cambridge: Cambridge University Press.

Rabinowicz, W. 1995. To have one's cake and eat it, too: sequential choice and expected-utility violations. *Journal of Philosophy* 92:586–620.

Weirich, P. 2010. *Collective Rationality: Equilibrium in Cooperative Games*. Oxford: Oxford University Press.

Weirich, P. 2015. *Models of Decision-Making: Simplifying Choices*. Cambridge: Cambridge University Press.

CHAPTER 4

Self-Control and Hyperbolic Discounting

Arif Ahmed

If self-control is lacking in a person – if he is *incontinent* – then he is consistently unable to overcome desires to get or do something. But this condition, though necessary, is not sufficient for incontinence as I understand it. Nobody can overcome his natural desire to breathe, to eat, or to sleep, but regularly doing all three is not incontinence. There must also be some sort of internal conflict: part of you wants to engage in the behavior and part wants to avoid it. Here I set out a model of this conflict as between *desires*.[1] The model is reasonably well known, but it is worth setting out how it captures three phenomena of the target state.

The first phenomenon is that the conflict would be resolved in different ways at different times. When a temptation is far off in the future, you are inclined not to succumb to it, but when the temptation is at hand, you *do* succumb. Your present desire to go running tomorrow morning certainly outweighs your present desire to stay in bed tomorrow morning, but tomorrow morning it is a different story. This *preference reversal* is a basic feature of incontinence that any model of it should capture.

The second phenomenon is that incontinence is more than the existence and eventual resolution of just *any* conflict between *any* desires. Before your morning run, you might be said to have a desire to go on your run and a desire to stay in bed. If anything should count as a conflict between competing desires, then so should this. But when the desire to go running wins out, it is not a failure but an *exercise* of self-control.

What is also necessary is that the desire that wins out must be more impulsive. You have a steady desire to go running every morning; last night you had the desire to go running tomorrow and no desire to stay in bed.

---

[1] This characterization of the target phenomenon is obviously different from, e.g., Davidson's classic definition of incontinence as intentionally doing something when you think that, all things considered, an available alternative is better (Davidson 2001 [1970]: 22). The aim here is not to give the correct analysis of a preexisting concept but rather to explicate self-control and incontinence so as to cover a range of interesting psychological phenomena.

This morning you have an impulsive desire to stay in bed. If the former wins out, you go running; if the latter wins out, then you stay in bed. What makes the latter case a failure of self-control is that here it is the ephemeral desire that wins out. The model should capture a distinction between ephemeral and steady desires.

The third phenomenon is that there is a genuine conflict between real desires: the component vectors, and not only their net resultant, must themselves enter into the description of the incontinent. This raises a particular difficulty for the attractively austere conception of desires, or rather preferences, that is standard in economic analysis. On this conception, the totality of facts about your desires is just the totality of facts about your preferences, where to prefer *A* to *B* just *is* to choose, or to be disposed to choose, *A* when *B* is the only alternative.[2] But if preference is just a fact about choices, then it is not clear how incontinence can be a fact about preferences, for a person's choices are simply the net resultant of her desires, and it is unclear how any actual choices, or dispositions to choose, can reveal conflict among the desires that gave rise to them.

Here I set out how the model of incontinence that I have in mind captures all three phenomena, before turning to two applications that are of interest in the contexts of decision theory and public policy.

4.1   The Hyperbolic Discounting Model

The model treats incontinence as a phenomenon of future discounting. To have it is to discount the future in ways that the continent do not. Making this clearer involves (1) saying what future discounting is, (2) distinguishing the forms it may take, and (3) setting out the descriptive consequences of one of these forms.

1. To discount the future is to value future less than present goods and distant future less than near-future goods, other things being equal. These "other things" include (a) one's confidence that one will get to consume the goods at the times they are promised, (b) the value of those goods to one's self *at* those times, and (c) the opportunity cost of foregoing either good. If these things are unequal, then it may carry little interest that one prefers the earlier to the later. So *future discounting*, as I'll understand it, covers as much of this preference as *survives* equalization of (a)–(c).

Suppose, e.g., that I now prefer $100 tomorrow to $100 in a year's time. (a) I might be significantly more confident of being alive tomorrow than of being

---

[2] See, e.g., Savage 1972: 17.

alive in a year, and money that I get when I am dead has less present value. (b) I might be confident that I will be richer in real terms in a year's time, so – given the declining marginal value of money – I have reason to expect $100 to be worth less to me in a year's time than it is now. (c) I might be confident that the nominal interest rate over the next year is positive, so I could invest the $100 now for a return next year that is strictly more than if I waited one year to get the $100. When I say that a person *discounts the future*, I mean that she prefers $100 or equivalent consumption now to $100 or equivalent consumption in the future, but not for any of the reasons (a)–(c).

We can describe the rate at which one does this in terms of a discount function f that maps time from the present (represented by a positive real number) to a real number between 0 and 1. $f(t)$ is the present dollar value of $1, or $1 of consumption at present prices, received at time t.[3] Thus suppose, e.g., that if time is measured in days, then $f(t) = 2^{-t}$. Then you would, today, pay up to 50 cents in exchange for (a contract promising) $1 to be paid tomorrow, up to 25 cents for $1 to be paid the day after tomorrow, and so on. The only constraints here assumed on the discount function are (a) that it satisfies $f(0) = 1$, reflecting the fact that you would pay up to $1 today in exchange for $1 today, (b) that it is nonincreasing, reflecting the fact that you do not value consumption in the more distant future at a higher rate than you value consumption in the less distant future, (c) that $f(t) > 0$, reflecting the fact that your value for a future good increases with the value that it would have for you if it were presently available.

2. Within the very wide range of discount functions that these constraints allow, we can distinguish *exponential* and *nonexponential* functions. Exponential discount functions – e.g., $f(t) = 2^{-t}$ – can always be written in the form $f(t) = e^{-rt}$ for some constant discount rate $r \geq 0$. The important feature of these functions is that *a fixed delay devalues a good by the same factor, whenever that delay occurs*. I call this feature *delay consistency*; we may define it formally as

Delay consistency: A discount function f is *delay consistent* iff:

$$\forall d, t_1, t_2 \geq 0 : \frac{f(t_1 + d)}{f(t_1)} = \frac{f(t_2 + d)}{f(t_2)} \tag{4.1}$$

This means that whether you'd now be willing to exchange the promise of a good at some future point for a larger good a day (a week, a year) later

---

[3] I am assuming for simplicity that money has constant marginal value; this will not create complications because we are ruling out (b).

than that point is independent of when that point is. It follows that if the exchange seems fair to you at any time, then it will continue to seem fair to you right up to the time at which the earlier good is due to be realized. It is easy to see that exponential discount functions are always delay consistent; given certain assumptions, it is also easy to show that delay-consistent discount functions are always exponential.[4]

Delay-consistent discount functions attach the same importance to a delay whenever it occurs. Within the class of discount functions that are *not* delay consistent, we distinguish those that treat a delay as *less* important the *further* it occurs in the future.

Impulsivity: A discount function *is impulsive* iff:

$$\forall d, t_1, t_2 \geq 0 : t_1 < t_2 \rightarrow \frac{f(t_1 + d)}{f(t_1)} < \frac{f(t_2 + d)}{f(t_2)} \quad (4.2)$$

This means that your willingness to exchange a good at some point in the future, for a larger good a day (a week, a year) later than that, *rises* as that point recedes into the future, so a delay that (now) seems unacceptable if it occurs tomorrow might (now) seem unproblematic if it occurs in a month. There is reason to think that humans discount the future impulsively: for instance, you would perhaps rather have one apple today than two tomorrow but also (now) rather have two apples in 51 days' time than one apple in 50.[5] A simple impulsive discount function is the hyperbolic form[6]

$$f(t) = \frac{1}{1 + kt} \quad k > 0$$

3. I propose to identify impulsivity of the discount function with incontinence, on the grounds that impulsivity generates the three phenomena: (a) preference reversal, (b) prioritization of ephemeral over stable desires, and (c) a sense in which there is a synchronic conflict between desires. These three phenomena arise as follows:

(a) Suppose that your discount function f takes the impulsive form (4.2) and that you now face the possibility of a reward of \$1 that is t_2 units of time in the future, together with a cost of \$$c$ that is $t_2 + d$ units of time in the

---

[4] Proof of the second claim: we know from (1) that $[f(t+d)]/[f(t)]$ is constant for any fixed d; hence we have $f(t)f'(t+d) - f'(t)f(t+d) = 0$ for any fixed d. So, for any d, t, we have $[f'(t)]/[f(t)] = [f'(t+d)]/[f(t+d)]$. Hence $\forall t f'(t) = -rf(t)$ for some constant $r \geq 0$, and the result follows. The assumptions are that f is differentiable and that it is strictly decreasing.

[5] For more evidence of this, see Green et al. 2005.

[6] Proof: Let $g(d, t) = [f(t+d)]/[f(t)]$. Then $\partial g/\partial t = k^2 d/(1 + kt + kd^2)$, so $d > 0 \rightarrow \forall t > 0 \ (\partial g/\partial t) > 0$.

future, where c has the property that for some t_1 such that $0 < t_1 < t_2$ we have

$$\frac{f(t_1 + d)}{f(t_1)} < \frac{1}{c} < \frac{f(t_2 + d)}{f(t_2)} \tag{4.3}$$

(It follows from (4.2) that some such c exists.) If you decline the reward, then your utility at all times is set to 0. Now, initially, i.e., when the reward is t_2 units in the future, you prefer to decline it because $f(t_2) < cf(t_2 + d)$. But, as time goes on, you eventually reach a time at which the reward is only t_1 units in the future. At this point you prefer to accept the reward because $cf(t_1 + d) < f(t_1)$. Your preferences have reversed.

Suppose, e.g., that you discount the future by $f(t) = 1/(1 + 4t)$, with t measured in days (i.e., 24-hour periods). Suppose that you know that tomorrow night you will choose whether to spend that evening drinking or sleeping. The pleasure of drinking will then be worth \$1 to you; the hangover, which inevitably occurs 12 hours later, will at that later time be worth –\$2 to you (so $c = -2$). Sleeping takes a benchmark value of \$0. Right now you are contemplating the prospect of either (a) drinking in 24 hours and a hangover in 36 hours or (b) not drinking and no hangover. From the present perspective, the value of drinking is $[1/(1 + 4.1)] - \{2/[1 + 4(1.5)]\} = -0.086$. So 24 hours before the moment of choice, you prefer not to drink; you may resolve not to drink. But as the moment gets closer and in consequence of the fact that f is sufficiently impulsive, you will reach a point at which that changes. By the time you get to the moment of choice, the value of drinking is $[1/(1 + 4.0)] - \{2/[1 + 4(0.5)]\} \approx 0.333$. At that point you prefer to drink, so at some point between then and 24 hours previously your preferences must have switched. This effect does not arise on the assumption that the discount function is delay consistent.

(b) Similarly, an impulsive discount function creates a prioritization of ephemeral over steady desires. Suppose that you are contemplating drinking on the Nth evening from now, tonight being the zeroth. The present value of drinking then is $f(N) - 2f(N + 0.5)$. For large enough N and a suitable discount function, e.g., $f(t) = 1/(1 + 4t)$, this present value is negative, so for all times up to a certain point you prefer not to drink on the Nth evening. But for a brief period up to and including the moment of decision, the present value of drinking on that evening is positive, so this ephemeral desire wins out over the long-lived desire not to drink on that evening. But, again, this phenomenon is inconsistent with any delay-consistent (i.e., exponential) discount function.

(c) By the same reasoning, impulsivity generates something like a synchronic conflict between short- and long-term goals because these are revealed in behavioral dispositions. I say "something like" because the conflict is not *direct* in the obvious and possibly incoherent sense that for objects of choice *A* and *B*, you both prefer *A* to *B* and prefer *B* to *A*. But, as we can see from (b), there is a discrepancy between what you prefer (and would bring about, if you could) for yourself in the distant future and what you prefer (and would bring about for yourself, if you could) in the short term. *Both* preferences, and not only the outcome of their resolution, are visible in the behavioral dispositions of the impulsive agent, such as, for instance, in St. Augustine's prayer that he be granted continence, only not yet.

4.2 Addiction as a Newcomb Problem

I turn now to two applications of the model. The first application is that incontinence gives rise to real-life versions of Newcomb's problem, that is, cases where the symptomatic value of an act diverges from its causal efficacy in specific ways that carry a certain kind of philosophical interest.

In order to explain this, I distinguish two theories of rational choice in the face of uncertainty: causal decision theory (CDT) and evidential decision theory (EDT). Informally, the difference is as follows: CDT says that when you have a choice between acts, you should do whatever you expect to have the best *effects*, whereas EDT says that you should do whatever brings the best *news*: the act that is the best evidence that things are or will be the way that you (now) want them to be.[7]

More formally, suppose that your beliefs can be represented as a *credence* function, i.e., a probability function Cr on a set (an algebra) of propositions, where we treat propositions as subsets of the set W of possible worlds such that $Cr(p) = 1$ represents full confidence in p, and $Cr(p) = 0$ represents full confidence in $\sim p = W - p$. The *subjective news value* V of any proposition satisfies the following constraint: if p is a nonempty proposition, and if the set $\{q_1, q_2, ..., q_n\}$ of propositions is a partition of W, then

$$V(p) = \sum_{i=1}^{n} V(p \cap q_i) Cr(q_i|p) \qquad (4.4)$$

---

[7] Jeffrey (1965) gives the standard exposition of EDT. [Note that Jeffrey (1983) amends the theory in the light of Newcomb's problem and in consequence effaces the difference from CDT that I wish to exploit in this section.] For a standard exposition of CDT, see Lewis (1981) or Joyce (1999).

In (4.4), $Cr(q_i|p)=_{\text{def.}} Cr(p \cap q_i)/Cr(p)$ is the *conditional probability* of q_i given p; it measures the extent to which the truth of p is, for you, evidence for the truth of q_i.

Now define the utility U of a proposition p as follows: let the partition $\{q_1, q_2, ..., q_n\}$ settle everything that matters to you, given that p is true.[8] Then we write

$$U(p) = \sum_{i=1}^{n} V(p \cap q_i) Cr(p \Rightarrow q_i) \qquad (4.5)$$

In (4.5), $p \Rightarrow q_i$ is the *subjunctive* or *counterfactual conditional* "If it *were* the case that p, then it *would be* the case that q_i." Your credence in this proposition measures the extent to which the truth of p causally promotes the truth of q_i, in your opinion.[9]

We now define EDT and CDT. If you face a choice between realizing the truth of propositions $p_1, p_2, ..., p_m$, then EDT says that you should realize any one that has *maximal news value* V from among the p_j, whereas CDT says that you should realize any one that has *maximal utility* U. This difference reflects the fact that EDT emphasizes news value, whereas CDT emphasizes efficacy, for the weights that each theory assigns to the value of an outcome $p \cap q_i$ of a given act p correspond to your estimates, respectively, of the degree to which p is evidence for and the degree to which it causally promotes that outcome.

This disagreement between EDT and CDT reflects a philosophical tension over the concepts of causality and related modal notions such as counterfactual dependence, as expressed here by $\Rightarrow$. Ever since Berkeley and Hume noticed that we never perceive a causal nexus between events but only the statistical frequencies with which they are associated, philosophers with empiricist sympathies have been skeptical about the existence of, or at any rate the point of believing in, any connection that goes beyond talk of such frequencies. In light of the further fact that causality appears to play no role in fundamental physics, some have gone so far as to deny the existence of causality altogether – most famously Russell, who wrote that "[t]he law of causality ... is a relic of a bygone age, surviving, like the monarchy, only because it is erroneously supposed to do no harm."[10] As for

---

[8] Formally, this means that $1 \le i \le n \rightarrow V(p \cap q_i \cap r) = V(p \cap q_i)$ for any nonempty $(p \cap q_i \cap r)$.

[9] This very simple treatment of utility follows that of Gibbard and Harper (1978). Technical reasons exist for preferring the more sophisticated treatment in Joyce (1999). But we may ignore them here.

[10] Berkeley 1980 (1710): part 1, sec. 25; Hume 1978 (1739): book 1, part 3; Russell 1986 (1913): 173.

related modal notions, Hume himself was as skeptical about the objectivity of any "metaphysical" necessity that supports such dependence as he was about objective causality itself, and in more recent times, Quine has expressed grave doubt that an empirically responsible scientific theory could say anything about the nonactual possibilities that subjunctive conditionals are supposed to be about.[11]

But, if CDT is right about what distinguishes rational from irrational actions, then we have reasons to believe in causality that are independent of whether perception or physics is sensitive to it: we need causal or counterfactual beliefs – specifically, credences in the $p \Rightarrow q_i$ – in order to make rational *choices*.[12] Of course, if CDT *were* right, our problems would not be over: we should still have to resolve the fundamental tension that would then seem to arise between the practical and theoretical constraints on belief, between what we need to believe to get by and how science tells us things really are, a tension that Field calls the "central problem in the metaphysics of causation."[13]

In contrast, if EDT is correct, then Hume, Russell, and Quine would seem to be doubly vindicated: causal beliefs are useless as well as theoretically unjustified. The philosophically crucial feature of EDT is precisely that its recommendations are wholly insensitive to the agent's specifically causal or counterfactual beliefs. The doxastic weights in equation (4.4) – i.e., of the form $Cr(q_i|p)$ – do not reflect causal judgments but are rather and indifferently registers of *any* sort of statistical association that creates expectations about the present case, whether or not any causal mechanism or counterfactual dependence underwrites the latter. It is therefore a matter of metaphysical interest whether we can find cases of practical decision making over which EDT and CDT disagree about what to do.

Now in a wide range of realistic decision-making scenarios these theories do *not* disagree, the reason being that in those cases the evidential import of a contemplated act is exhausted by its evidential bearing on its *effects*. For instance, it might seem that EDT advises one not to visit the doctor because doing so is evidence – but does nothing to bring it about – that one is ill.[14] However it is unlikely that you take your own currently contemplated visit as evidence that you are ill (although *somebody else* might take it as evidence that you are ill). After all, while you know that the incidence of illness among people found in doctors' offices is higher than in the general population, you also know that among people with

[11] Hume 1978 (1739): book 1, part 3, chap. 14; Quine 1975.   [12] Cartwright 1979.
[13] Field 2004: 443.   [14] Skyrms 1980: 130.

your symptoms, the incidence of illness is the same among people who do and people who do not visit the doctor. So, for you, and more generally for anyone who knows what your symptoms are, visiting the doctor neither brings about *nor is evidence of* your being ill: EDT and CDT therefore recommend such a visit in exactly the same circumstances. The same can be said for many other cases where EDT and CDT seem to disagree.[15] And the best-known case where they certainly do diverge – the original version of the Newcomb problem – was explicitly introduced into the philosophical literature as a science-fictional case.[16]

Returning now to the impulsivity model of incontinence, an interesting consequence of that model is that there are much more plausible scenarios in which EDT and CDT do make conflicting recommendations. Here is an intuitive description of such a case. Suppose that you are facing an indefinite sequence of evenings on which you will get the opportunity to drink and that you think, as might be reasonable, that what you choose tonight will be evidence of, but will not have any effect on, what you do on later evenings. More specifically, we can suppose that drinking tonight is evidence of, although it does nothing to shape, a taste for alcohol. Then, if you are keen to drink tonight but especially keen not to drink in the more distant future, EDT might advise you not to drink tonight because doing so would be evidence of future indulgence, whereas CDT would advise you to go ahead and drink tonight because your drinking tonight *makes* no difference to your drinking tomorrow.[17]

Here is a crude way to make that picture more definite: suppose that (a) your discount function is $f = f(t)$, (b) the choice on any evening is between one drink and none, (c) the utility of a drink is $1 enjoyed at the time of consumption and the cost is $c paid d units of time later (where the unit of time is 1 day), (d) if you drink tonight, then your expected number of drinks per evening on all future occasions is m, and (e) if you do not drink tonight, then your expected number of drinks per evening on all future occasions is $n < m$. (f) Write $F(t)$ for the present discounted value of a drink t days in the future so that

[15] These include several allegedly realistic versions of Newcomb's problem, such as *medical* Newcomb problems (Lewis 1981: 310–11), *economic* Newcomb problems (Broome 1989), and *psychological* Newcomb problems (Shafir and Tversky 1992); for discussion of all of these, see Ahmed 2014: secs. 4.2, 4.4, and 4.5.
[16] Nozick 1969: 114*ff*.
[17] For a report on an experimental realization of this type of situation, see Kirby and Guastello (2001).

$$F(t) =_{\text{def.}} f(t) - cf(t+d) \qquad (4.6)$$

Finally, suppose that (g) you like a drink on any evening considered in isolation, i.e., $F(0) > 0$.

If we are given all these assumptions, then EDT and CDT make different recommendations if and only if the following condition holds:

$$F(0) + \sum_{t=1}^{\infty} (m - n)F(t) < 0 \qquad (4.7)$$

(For the proof, see Appendix 4A.[18]) Inequality (4.7) says that your drinking tonight makes enough of a difference to the probability of your drinking on future evenings for it to be worth your abstaining tonight (i.e., according to EDT). If conditions (a)–(g) are satisfied, then we have a real-life Newcomb problem in which it follows from (g) that CDT recommends giving in to temptation tonight, and it follows from (4.7) that EDT recommends that you resist it, at least for tonight.

Notice that a situation of this type can only arise if f is delay inconsistent: if f is exponential, then assumption (g) is inconsistent with (4.7). In contrast, it is easy to see that the situation does arise for at least some impulsive f: if, for instance, $f(t) = 1/(1+t)^2$, then there are values of c and d that jointly satisfy (a)–(g) as well as (4.7).[19]

One limitation of this result is that the difference between EDT and CDT is unlikely to persist in the face of accumulated experience of one's own choices, *if* each past choice is equally relevant to one's estimate of one's own future drinking. As you experience successive evenings on which you either choose to drink or choose not to drink, your beliefs about your own

[18] In writing this inequality, I suppose that the horizon is infinite, i.e., that the evenings on which you get the opportunity to drink go on forever. This is unrealistic in a way that might seem to matter for my argument. After all, if you know that the Nth evening is your last opportunity to drink, then a backward induction argument would seem to recommend drinking on every evening regardless of whether EDT or CDT is the background decision theory. One possible response is that treating the problem as infinite is simply a device for dramatizing your ignorance as to which evening is the last drinking opportunity and that on each evening there is a fixed probability p that you will get another chance tomorrow (see, e.g., Gibbons 1992: 90). The consequent refinement of the model would not affect the main qualitative results derived here.

[19] Putting $c = 3$ and $d = 1$ *gives* $F(0) = 0.25$. Condition (g) is therefore satisfied. Also in this case we have $\sum_{t=1}^{\infty} F(t) = (33 - 4\pi^2)/12$, so (4.7) is satisfied if $m - n > 3/(4\pi^2 - 33)$. Notice that the straightforward hyperbolic function $f(t) = 1/(1 + kt)$ can also meet these conditions, but only in a degenerate way, because for that discount function the sum $\sum_{t-1}^{T} F(t)$ diverges as $T \to \infty$ if $c > 1$. The refinement mentioned in note 18 would avoid this degeneracy while still implying (for large enough p) that EDT recommends abstention tonight.

tastes will become set: your decision on each subsequent evening will then become less and less evidentially relevant to whether you drink on any future evening. For an extreme example of this *smoothing*, if I have been drinking on every evening for the past 10 years, then my not drinking tonight is unlikely to be much evidence that I am going to drink less in future: the probability of drinking on any future evening, given that I drink tonight or given that I do not drink tonight, is in either case high. More formally, as time goes on, we should expect the values of m and n to converge and hence that $m - n$ should fall to 0. So there will inevitably come an evening on which, and after which, (4.7) fails: from that point on, EDT and CDT agree that you should drink, and the Newcomb-like character of the problem is lost. (For a formal elaboration of the model that gives rise to this result, see Appendix 4B.)

In contrast, if we give up on the (implausible) idea that each past choice is evidentially equally relevant to the probability of drinking on any future occasion, then we can construct other models in which EDT steadily recommends abstention (and CDT steadily recommends drinking). In these models, whether you drink *tonight* has more weight than whether you drank (say) on this evening last year in settling your probability of drinking on future occasions. In the limit, we can suppose that your drinking behavior is, in your own steady opinion, a Markov process: the probability of your drinking on any evening is a function of whether you drank the previous evening.[20]

If we now write m (respectively, n) for the probability that you drink on any night given that you drink on the preceding night (respectively, given that you do not drink the preceding night), then the critical condition is

$$F(0) + \sum_{t=1}^{\infty} (m - n)^t F(t) < 0 \qquad (4.8)$$

If (4.8) holds, then EDT will recommend abstention on every night; if in addition $F(0) > 0$, then CDT will recommend drinking on every night. (For proof, see Appendix 4C.) Equation (4.8) is inconsistent with $F(0) > 0$ if discounting is exponential, but there certainly are impulsive discount functions that satisfy both conditions.

In general, we can say this: if your expectations about future behavior given past behavior weights recent data heavily enough, then repeated temptation puts sufficiently incontinent subjects in the position of facing

---

[20] This simplification follows von Weiszäcker (1971).

a repeated Newcomb problem in which EDT and CDT diverge. The follower of EDT will always resist temptation; the follower of CDT will always give in to it. Far from illustrating the pragmatic *need* for beliefs about causal dependence and independence, incontinence is therefore a relatively realistic illustration of how such beliefs can be counterproductive. In so far as it supports any metaphysical position in this dispute, what it supports is the relatively severe empiricism of Hume or Russell or Quine.[21]

4.3 Incontinence and Political Intervention

I turn now to a more practical application of the model, namely the evaluation of political and economic devices aimed at addressing incontinence. Suppose that there are forms of addiction for which the foregoing model is correct, and suppose that the state wishes to eliminate or reduce the occurrence of this form of addiction.[22] How should it do this?

In her recent book on paternalism, Conly (2013) identifies three broad types of approaches. First, there is the *liberal* strategy: place no restrictions on the individual's behavior, but instead create a "better" environment in which those decisions can take place. Second, there is the *libertarian paternalist* strategy (*nudging*): present their choices in ways that exploit people's biases – e.g., by controlling what gets to count as the *default option*. Third, there is the *coercive* strategy (which Conly generally favors): simply ban the behavior, e.g., smoking and drinking.

I won't discuss the second and third options here, not because they are obviously wrong – there is much to be said for both – but because I myself value individual liberty enough to think that if the state cannot restrict addictive consumption through liberal methods, then it ought not to restrict it at all. To this end, I'll consider the liberal options that Conly discusses before arguing that if incontinence is a matter of impulsive discounting, then there are liberal responses to it that may avoid her criticisms.

---

[21] For further discussion of this type of "bundling," see Ainslie (1991). In that paper, Ainslie does not consider EDT as a bundling mechanism; for a brief discussion of evidential considerations such as those discussed here, see Ainslie (2012: 20), which, however, restricts the discussion to pure hyperbolic discounting. The present analysis applies to *any* impulsive discount function, which class includes but is not exhausted by hyperbolic discount functions. [For instance, $f(t) = (1 + kt)^{-n}$ is impulsive for any $n > 0$ but is only hyperbolic when $n = 1$.]

[22] It may have various reasons for wanting to do that, but I won't go into these – see, e.g., Foddy and Savulescu (2010) on (what political actors often take to be) the costs of addiction to oneself; see, e.g., Jonson et al. (2012) on the perceived social costs of gambling addiction in the state of Victoria.

Conly mentions two such options: "education" and "experience." In the case of addiction, the first involves teaching people either (a) about the dangers involved in consumptions, for instance, by forcing manufacturers to put warning labels on products, or (b) to avoid the cognitive biases that create addiction in the first place. She argues against (a) that it has proven unsuccessful, for instance, with regard to smoking.

> It's true that a smaller percentage of the population smokes now than did before it was discovered that smoking causes cancer. On the other hand, more than 20 percent of the American population does smoke, despite the millions of dollars spent in schools and the unmissable warnings of cigarette packages ... Educating people out of error is not easy, when errors arise from cognitive bias.[23]

Against (b), she argues similarly that educating people out of cognitive biases is likely to be much more complicated than simply imparting the facts and that if such education is to be effectively applied, individuals must be able to tell when they are succumbing to this or that cognitive bias, and they cannot reliably do that.[24]

The second liberal option, *experience*, is not any sort of intervention but rather a principled refusal to intervene, the principle being that people learn better from making their own mistakes than through formal education. Conly's response is that typically such learning either doesn't happen at all (consider the persistence with which many highly educated people procrastinate) or happens too late to do any good.

> When you are forced to retire and find out that you don't have enough to live on, it's too late to learn that your tendency toward irrational time discounting has really done you harm.[25]

Both liberal options presuppose that the problem is intellectual: there is something that the agent doesn't know or knows but doesn't bear in mind. The idea is, e.g., that the only reason, or the main reason, that people, e.g., smoke is that they are unaware of or inattentive to the danger of smoking, and the reason they don't save enough is that they are unaware of, or do not attend carefully enough to, the effect that this will have further down the line.

This picture is at odds with the model of incontinence presented here: an impulsive discounting function is entirely consistent with being aware of the costs of the impulsive behavior. You might know perfectly well that if you drink tonight, then your hangover tomorrow will make you regret it;

---

[23] Conly 2013: 25.   [24] Conly 2013: 25–26.   [25] Conly 2013: 27.

still, if your discount function takes the form described, then you are going to drink anyway. The problem isn't that you are ignorant about what the consequences are like – you may know from long experience exactly what a hangover will be like – but that you just don't care, or rather you don't care *enough*. But not caring "enough" is neither obviously mistaken nor obviously irrational: it just means that at that time you care less than the time at which you are assessing the decision to drink, either long before the decision or during the subsequent hangover.

To say, as I just said, that impulsive discounting is not *irrational* is a roughly Humean view.[26] But to say – as I also think – that it is not otherwise *mistaken* appears vulnerable to a Humean objection, namely that such discounting – perhaps indeed *any* sort of discounting – is tied up with a failure to appreciate with full *vividness* just what the harms of the contemplated behavior really are, when those harms occur far enough in the future.[27]

But let it be true that our imagination makes the future harms (and perhaps also goods) that are more distant in time less vivid than those that are near. It doesn't follow that the imagination in any way *distorts* the "real and intrinsic value" of those goods, as Hume calls it. All it shows is that it makes a great difference to our contemplation of some future event whether that event is supposed to be close or distant. But why think that from only one of these perspectives do we perceive the event's true value? Even if there is such a thing as its true value, perhaps it, like futurity itself, is irreducibly tensed. In any event, there is no straightforward argument, from the fact that the imagination presents different times in different lights, to the conclusion that this involves any sort of mistake. To think so

---

[26] "Where a passion is neither founded on false suppositions, nor chooses means insufficient for the end, the understanding can neither justify nor condemn it. It is not contrary to reason to prefer the destruction of the whole world to the scratching of my finger. It is not contrary to reason for me to choose my total ruin, to prevent the least uneasiness of an Indian or person wholly unknown to me" (Hume 1978 (1739): book 2, part 2, chap. 3). It is worth adding here that even if impulsive discounting is irrational, there is plenty of evidence that it is descriptively plausible (see, e.g., Ainslie 2001), and so the consequences of it that are drawn here (and the much more extensive deductions that are available in Ainslie's work) would still be of interest.

[27] "What strikes upon [us] with a strong and lively idea commonly prevails above what lies in a more obscure light; and it must be a great superiority of value, that is able to compensate this advantage. Now as every thing, that is contiguous to us, either in space or time, strikes upon us with such an idea, it has a proportional effect on the will and passions, and commonly operates with more force than any object, that lies in a more distant and obscure light. Though we may be fully convinced, that the latter object excels the former, we are not able to regulate our actions by this judgment; but yield to the solicitations of our passions, which always plead in favour of whatever is near and contiguous. This is the reason why men so often act in contradiction to their known interest" (Hume 1978 (1739): book 3, part 2, chap. 7).

would be comparable to inferring, from the premise that an object has a different visual appearance when viewed from different distances, to the conclusion that from every distance, or from every distance except one, we fail to see it as it really is.[28]

4.4 Impulsivity and Precommitment

I want now to consider an alternative liberal approach that does not presuppose that impulsive discounting is irrational or mistaken but rather simply exploits the preferences that that form of discounting itself creates. Someone who discounts the future impulsively takes different attitudes toward a delay of fixed length when that delay is distant as compared to when it is imminent. So, if an activity has an immediate reward and a delayed penalty, that person may take different attitudes toward the activity when it is distant and when it is imminent. Whether this happens depends on the relative size of the reward and the penalty and the length of the delay. With the discount function f, the penalty-to-reward ratio c, and delay d, the general condition for a clash is that there be positive real numbers t_1 and t_2 satisfying equation (4.3).

We have also seen that this feature of incontinence, that one willingly does something that one both foresees and regrets, or "pregrets," is a consequence neither of future discounting by itself nor of future discounting at any level of steepness. If your $f(t) = e^{-rt}$, then you will indeed discount the future, and as r rises, your behavior will sink into the most wanton hedonism, but you will not regret anything in virtue of this. You are not then incontinent but rather intemperate.[29]

---

[28] One might claim that impulsive discounting (though not other kinds of discounting) arises from a different cognitive error: that of not conceiving of equal temporal intervals as equal but rather as longer or shorter depending on their distance from the present. Thus it has been conjectured that our conception of both time and value follow Weber's law governing perception (Gibbon 1977), according to which the perceived magnitude of an increment is inversely proportional to the quantity of value or time that it is an increment *to*. Applying this form of "distortion" to a delay-consistent discount function may give rise to an impulsive one (Cui 2011). Putting it crudely, we discount impulsively because a delay of 12 hours one year from now looks now like less of a delay than does a delay of 12 hours one minute from now. But again it is not clear that this kind of distortion involves anything like a mistake. Nobody *believes* the false proposition that the extent of a given temporal interval really *is* less if it occurs further in the future as is clear from the fact that nobody makes this kind of mistake when calculating, e.g., the return on a bond. So what sort of "mistake" is involved here? Besides, even if this proposal were the right explanation of impulsive discounting, it would not follow that impulsivity *itself* involves a mistake but only that it is the (possibly happy) consequence of one.

[29] "One person pursues excess of unpleasant things because they are excesses and because he decides on it, for themselves and not for some further result. He is intemperate; for he is bound to have no regrets, and is incurable, since someone without regrets is incurable" (Aristotle 1985: 1150a18–24).

But since impulsive discounting *does* involve behavior that is regretted if foreseen, the impulsive discounter (and only he) is willing to bind his future self. He is willing to give up some present consumption in exchange for the foreclosing of certain options to his own future self. To get a measure of how much he will give up, consider a one-off version of the drinking example from Section 4.2. (From now on we can focus exclusively on the one-off case.) Given a discount function $f(t) = 1/(1 + 4t)$, a reward to drinking of \$1 to be enjoyed at the time of consumption, and a cost to drinking of \$2 to be suffered 12 hours later, the value to you of drinking tomorrow is $\$([1/(1 + 4)] - \{2/[1 + 4(1.5)]\}) \approx -\$0.086$. So you are willing to pay about 9 cents now to bind yourself not to drink in 24 hours' time. For instance, suppose that a pill makes you incapable of drinking tomorrow evening if you take it now (with no side effects). You will pay up to 9 cents for the pill. That is, you will voluntarily and rationally finance the binding of your own future actions.

So impulsivity, like steep exponential discounting, makes people act in short-sighted ways, but also and unlike the latter, it motivates them to prevent themselves from acting in these ways. This is the element of truth in Hume's remark that "this infirmity of human nature becomes a remedy to *itself* . . . we provide against our negligence about remote objects, merely because we are naturally inclined to that negligence."[30] Hume himself saw in this the origin of government, i.e., of the fact that people voluntarily surrender their own liberty to Leviathan; be that as it may, it matters here because it suggests a third liberal response to incontinence.

Suppose that a liquor store offers, for a fee, the following service: a customer can have her ID recorded, and the store will undertake not to serve her for a fixed period, perhaps in perpetuity, starting at a fixed future date. The bearer of impulsive preferences will pay for this to happen if the date and the price are right. If the marginal costs of producing this good do not exceed the marginal willingness of enough customers to pay for it, we can expect a market to arise for this service. And if it does arise, it will mitigate the effects of impulsive discounting. A customer who is willing to purchase this service will bind herself not to buy drink from this store for some nominated period, and as that period approaches its end, we can expect the very same motivations that drove the customer to take up the offer to drive her to renew it.

---

[30] Hume 1978 (1739): book 3, part 3, chap. 7 (emphasis added).

Of course, prohibiting yourself from buying liquor from one store is not the same thing as prohibiting yourself from drinking. There may be other stores, or you may be able to get your friends to stand you a drink. Self-binding can never eliminate but can only raise the *cost* of drinking. In this respect, though, it is no different from taxation, regulation, or even prohibition. Anyway, as long as the cost exceeds your future marginal willingness to pay for liquor, self-binding will be effective. What this cost is will vary depending on the context. For instance, if you live in a small and isolated town in which only one retailer sells liquor, then it may be true – and you may be able to foresee – that your desire for a drink in the future will not outweigh the inconvenience and opportunity cost of driving to the next town to get one or the awkwardness of scrounging yet again off your friends. Here a contract of the simple sort envisioned is likely to be effective.

In contrast, if you live in a city where alcohol is widely available, it may be necessary to make this sort of contract with all retailers, but the overheads associated with each contract might mean that the price of self-prohibition from retailer A does not outweigh, for any individual, the marginal benefits of not being able in the future to purchase alcohol from *that* retailer. One possible solution would be for all retailers to agree to offer a single contract. A more paternalistic solution may also be available: the state could make it a requirement of holding a license to sell alcohol that the retailer offers self-prohibiting contracts at or below a certain maximum price, thus transferring some or all of the costs of self-prohibition from the consumer to the supplier. In many countries, this solution could exploit the fact that an infrastructure for the distribution of licenses, and presumably also for the supervision of licensees, already exists.

For an example of how this might work in practice, consider the precommitment scheme for gambling that the Australian state of Victoria has required casinos to offer since December 2015 ("YourPlay").[31] Registration (which is not compulsory) entitles the gambler to a card on which he can record activity (plays, wins, and losses) and with which he can preset limits to time spent gambling, or to net losses incurred, for some future period. These limits can be changed at any time up to the start of play, either online or at specially provided kiosks at the gambling venues; they cannot be changed during play itself.

---

[31] For details of the scheme, see www.yourplay.com.au. For details of the regulatory framework, see Gambling Regulation Amendment (Pre-commitment) Act 2014, available at www.legislation.vic .gov.au (accessed July 19, 2017).

The legislation underlying this scheme grew out of a recommendation of a 2010 report by the Australian Productivity Commission.[32] At the time of its publication, casinos protested that the proposals were paternalistic and that they would reduce both gambling revenues and the consequent benefits to communities in which casinos operated.[33] Both concerns are in my view reasonable, but the first certainly invites qualification. Paternalism can hardly consist of the making available of *additional* options to individual gamblers. It is true that the state is interfering with the free operation of the *market*. It isn't so clear that it is interfering with the liberty of the *individual consumer* in any way that liberals (by which I mean classical liberals or libertarians) should find equally troubling.

The foregoing analysis suggests two possible improvements to this approach, one from a political perspective and the other from an economic perspective, proceeding on the assumption that addiction is a phenomenon of impulsive discounting. The first is that the option to precommit need not be *imposed* on casinos. If gamblers want to precommit, then they will pay to precommit. So casinos should be willing to provide this option in advance. A system of this sort would help address the concern that the proposals are paternalistic, not in the sense of limiting consumer choice but in the sense of imposing all costs of a precommitment service on its suppliers rather than on its consumers. Still, a market for precommitment will arise only if it is economically viable, that is, if the marginal cost of providing it does not exceed the marginal willingness to pay for it. For a consumer who only slightly regrets her foreseeable future inclination to gamble (or to drink), the state might have to offer to cover the cost of precommitment, i.e., to pass it on to taxpayers. Or the state might mandate gambling operators to offer precommitment devices at no more than a fixed price (or for free).

The second possible improvement arises from the fact that if discounting is impulsive, then the optimal point at which to offer the option to precommit – the time at which that option has most value for the prospective gambler – is unlikely to be *immediately* prior to the commencement of the activity. When the opportunity to gamble is in the distant future, gambling will look like a bad prospect. But because it is in the *distant* future, it won't look like a *very* bad prospect. But, as the time of opportunity approaches, there are two competing effects. On the one hand, and because f is decreasing, the opportunity will start to matter more, and this will make gambling look like a *worse* prospect.

[32] Productivity Commission 2010.   [33] For an account of the debate, see Jonson et al. (2012).

On the other hand, and because f is impulsive, the delayed penalty of gambling (expected loss, feelings of guilt, etc.) will start to matter less relative to the reward (the excitement of the activity, etc.), and this will make gambling seem a *better* prospect. If initially the first effect dominates the second and later the second effect dominates the first, then there will be some time strictly *before* the point of opportunity at which gambling looks maximally undesirable. This will be the point at which one is maximally willing, and so willing to pay the most, to precommit.

Suppose, for instance, that $f(t) = 1/(1 + kt)$, $k > 0$, and that at some point in the future one will have an opportunity to gamble, with an immediate reward of 1 unit and a penalty of $c > 1$ after a delay of d. Then taking up the opportunity of gambling looks worst when its distance in the future is $t = t^*$ satisfying

$$t^* = \frac{d}{\sqrt{c} - 1} - \frac{1}{k} \tag{4.9}$$

(For proof, see Appendix 4D.) If this quantity is positive, then there will be some time *before* the opportunity to gamble at which the consumer's marginal willingness to pay for precommitment is maximal.

What this means in practical terms is that options to precommit ought to be available at least t^* units of time in advance of the opportunity that one is binding oneself against. A casino, for example, might offer customers the opportunity to ban themselves for a 24-hour period *starting in a week's time* (or some other delay depending on one's estimate of c, d, and k). If the hyperbolic model is correct, then offering this option would maximize both uptake of precommitment at any given price and the price that would sustain any given level of demand. Some existing precommitment mechanisms do incorporate some such feature. For instance, the YourPlay scheme prevents customers from *increasing* their limits on loss or time spent gambling in the 24 hours before play begins. Similarly, some US states impose "cooling-off" periods (e.g., 10 days in California, 3 days in Florida) between the time at which you purchase a firearm and the time at which you collect it. Although I have no evidence that impulsive discounting played a role in the formation of these regulations, it is an advantage of the model that it (a) rationalizes this approach and (b) suggests by means of (4.9) how it might be optimized.

4.5 Conclusion

I have argued that modeling incontinence as a phenomenon of impulsive discounting has decision-theoretic and political consequences. In decision theory, it makes room for an interpretation of Newcomb's problem that casts doubt on the presumption that it illustrates the practical advantage of causal beliefs. Politically, it makes visible a procedure by which government might mitigate the effects of incontinence without interference with the individual consumer of "incontinent goods" and by which the market might do so without imposing costs on the firms that supply them.

APPENDIX 4A

Proof that (4.7) Is a Necessary and Sufficient Condition for EDT and CDT to Give Different Recommendations

Suppose that there are $N + 1$ evenings on which you can drink, starting from tonight. Let P_i be the proposition that you take a drink on the ith evening, where tonight is the zeroth evening, and let $\sim P_i$ be the proposition that you do not. Let $N = \{1, 2, ..., N\}$, and define

$$J = \left\{ \bigcap_{k \in K} P_k \cap \bigcap_{h \in N-K} \sim P_h | K \subseteq N \right\}$$

J is the set of all possible distributions of drinking and nondrinking over the next N evenings starting from tomorrow. Clearly J is a partition.

By (4.4),

$$V(P_0) = \sum_{j \in J} Cr(j|P_0)V(j \cap P_0)$$

But

$$V(j \cap P_0) = F(0) + \sum_{t \in j} F(t)$$

So

$$V(P_0) = F(0) + \sum_{j \in J} Cr(j|P_0) \sum_{t \in j} F(t)$$

And

$$\sum_{j\in J} Cr(j|P_o) \sum_{t\in j} F(t) = \sum_{t\in N} F(t) \sum_{\substack{j\in J \\ j\subseteq P_t}} Cr(j|P_o)$$

$$= \sum_{t\in N} F(t) Cr(P_t|P_o) = \sum_{t\in N} mF(t)$$

So

$$V(P_o) = F(o) + \sum_{t\in N} mF(t)$$

So, as $N \to \infty$,

$$V(P_o) \to F(o) + \sum_{t=1}^{\infty} mF(t)$$

if that limit exists. Similarly,

$$V(P_o) = \sum_{t=1}^{\infty} nF(t)$$

EDT recommends not drinking tonight iff $V(P_o) < V(\sim P_o)$, hence iff $F(o) + \sum_{t=1}^{\infty}(m-n)\sum F(t) < o$. In contrast, by (4.5) and the causal independence of future episodes of drinking on the present, $U(P_o) = \sum_{j\in J} Cr(j) V(j\cap P_o)$ and $U(\sim P_o) = \sum_{j\in J} Cr(j)$ $V(j\cap \sim P_o)$. So $U(P_o) = F(o) + U(\sim P_o)$. Since $F(o) > o$, it follows that EDT and CDT give different recommendations if and only if (4.7) is true.

APPENDIX 4B

A Model on which EDT and CDT Make Convergent Recommendations

Suppose that you face indefinitely many opportunities to drink and that there is a fixed but unknown number P, a function of your (unknown) tastes, that represents your chance of drinking on any future evening. Suppose that you start out with a uniform distribution for P. Write Cr_n^y for your beliefs after y nights of drinking and n nights of abstention; write P_M for the proposition that you drink on the Mth evening. Then we can derive expressions for the evolution of your belief about your chance of drinking on any night and for the expected future value $V_n^y(P_{y+n+1})$ given that you do and $V_n^y(\sim P_{y+n+1})$ that you do not drink on a night following y nights of drinking and n nights of abstention as follows.

The probability of y nights of drinking and n nights of abstention given a chance x of drinking on one evening is

$$Cr(y, n | P = x) = \frac{(y + n)! x^y (1 - x)^n}{y! n!}$$

So

$$Cr_n^y(P = x) = \frac{x^y (1 - x)^n}{\displaystyle\int_0^1 z^y (1 - z)^n dz} = \frac{(y + n + 1)! x^y (1 - x)^n}{y! n!}$$

Hence

$$V_n^y(P_{y+n+1}) = F(0) + \int_0^1 Cr_n^{y+1}(P = x) \sum_{t=1}^{\infty} x F(t) dx$$

$$= F(0) + \frac{y + 2}{y + n + 3} \sum_{t=1}^{\infty} F(t)$$

and

$$V_n^y(\sim P_{y+n+1}) = \int_0^1 Cr_{n+1}^y(P = x) \sum_{t=1}^{\infty} x F(t) dx = \frac{y + 1}{y + n + 3} \sum_{t=1}^{\infty} F(t)$$

Therefore, EDT recommends on the Mth evening that you not drink iff the following condition holds:

$$-\frac{1}{M + 2} \sum_{t=1}^{\infty} F(t) > F(0)$$

If you like a drink "for itself" (i.e., if $F(0) > 0$), then this condition is never satisfied given a delay-consistent discount function f. It *may* be satisfied when M is suitably small and f is impulsive. But as M increases, the condition will eventually fail whatever your drinking history, so at some point EDT will start recommending that you drink and will continue to make that recommendation. In contrast, if U_n^y is your utility function after n nights of abstention and y nights of drinking, it is easy to see that

$$U_n^y(P_{y+n+1}) = F(o) + U_n^y(\sim P_{y+n+1})$$

so $F(o) > o$ implies that after enough evenings, CDT and EDT both recommend drinking on every evening.

APPENDIX 4C

Proof that on the Markov Process Model, (4.8) Is Sufficient for EDT to Recommend Abstention Every Evening

Suppose that your drinking choice on any night has evidential bearing on your drinking choice on the next night that swamps any evidence from your drinking behavior on preceding nights. Write P_t for the proposition that you drink t days from now. Then the basic assumption is that there are quantities m and n, with $o < n < m < 1$, satisfying $Cr(P_t|P_{t-1}) = m$ and $Cr(P_t|\sim P_{t-1}) = n$.

Writing $\sigma_t =_{\text{def.}} Cr^+(P_t)$ and $\pi_t =_{\text{def.}} Cr^-(P_t)$, we have, for $t > 1$,

$$\sigma_t = m\sigma_{t-1} + n(1 - \sigma_{t-1}) = (m-n)\sigma_{t-1} + n$$

and

$$\pi_t = m\pi_{t-1} + n(1 - \pi_{t-1}) = (m-n)\pi_{t-1} + n$$

Solving these difference equations, we have

$$\sigma_t = \sigma_1(m-n)^{(t-1)} + n\sum_{i=0}^{t-2}(m-n)^i = m(m-n)^{(t-1)} + n\sum_{i=0}^{t-2}(m-n)^i$$

$$\pi_t = \pi_1(m-n)^{(t-1)} + n\sum_{i=0}^{t-2}(m-n)^i = n(m-n)^{(t-1)} + n\sum_{i=0}^{t-2}(m-n)^i$$

But

$$V(P_o) = F(o) + \sum_{t=1}^{\infty}\sigma_t F(t)$$

and

$$V(\sim P_o) = \sum_{t=1}^{\infty}\pi_t F(t)$$

Substitution of the foregoing expressions for σ_t and π_t gives

$$V(\sim P_o) > V(P_o) \leftrightarrow F(o) + \sum_{t=1}^{\infty}F(t)(m-n)^t < o$$

APPENDIX 4D
Proof of Equation (4.9)

At $t = t^*$, $F'(t) = 0$, so $f'(t^*) = cf'(t^* + d)$, where $f(t) = 1/(1 + kt)$.
So $c(1 + t^*k)^2 = (1 + t^*k + dk)^2$; that is, $cu^2 = u^2 + 2udk + d^2k^2$ where
$u = 1 + t^*k$. Hence

$$u = \frac{2dk \pm \sqrt{4d^2k^2 + 4d^2k^2(c - 1)}}{2(c - 1)} = \frac{dk(1 \pm \sqrt{c})}{(c - 1)}$$

Now $c > 1$, so

$$t^* > 0 \rightarrow u = \frac{dk}{\sqrt{c} - 1}$$

Hence

$$t^* = \frac{d}{\sqrt{c} - 1} - \frac{1}{k}$$

It is easy to see that if t and d are both positive, then

$$\frac{k}{(1 + tk)^2} = \frac{ck}{(1 + tk + dk)^2} \rightarrow \frac{ck^2}{(1 + tk)^3} > \frac{k^2}{(1 + tk + dk)^3}$$

So $F''(t^*) > 0$ and t^* minimizes F.

References

Ahmed, A. 2014. *Evidence, Decision and Causality*. Cambridge: Cambridge University Press.

Ainslie, G. 1991. Derivation of "rational" economic behavior from hyperbolic discount curves. *American Economic Review* 81:334–40.

Ainslie, G. 2001. *Breakdown of Will*. Cambridge: Cambridge University Press.

Ainslie, G. 2012. Pure hyperbolic discount curves predict "eyes open" self-control. *Theory and Decision* 73:3–34.

Aristotle 1985. *Nicomachean Ethics*, trans. T. Irwin. Indianapolis, IN: Hackett.

Berkeley, G. 1980 (1710). *Principles of Human Knowledge*. In *Philosophical Works*, ed. M. Ayers. London: Everyman.

Broome, J. 1989. An economic Newcomb problem. *Analysis* 49:220–22.

Cartwright, N. 1979. Causal laws and effective strategies. *Noûs* 13:419–37.

Conly, S. 2013. *Against Autonomy*. Cambridge: Cambridge University Press.

Cui, X. 2011. Hyperbolic discounting emerges from the scalar property of interval timing. *Frontiers in Integrative Neuroscience* 5:24. doi: 10.3389/fnint.2011.00024

Davidson, D. 2001 (1970). How is weakness of the will possible? In *Essays on Actions and Events* (pp. 21–42). Oxford: Oxford University Press.

Field, H. 2004. Causation in a physical world. In *Oxford Handbook of Metaphysics*, ed. M. Loux and D. Zimmerman. Oxford: Oxford University Press.

Foddy, B., and J. Savulescu. 2010. A liberal account of addiction. *Philosophy, Psychiatry and Psychology* 17:1–22.

Gibbard, A., and W. Harper. 1978. Counterfactuals and two kinds of expected utility. In *Foundations and Applications of Decision Theory*, ed. C. Hooker, J. Leach, and E. McClennen (pp. 125–62). Dordrecht: Riedel.

Gibbon, J. 1977. Scalar expectancy theory and Weber's law in animal timing. *Psychological Review* 84:279–325.

Gibbons, R. 1992. *A Primer in Game Theory*. Harlow: Prentice-Hall.

Green, L., J. Myerson, and E. W. Macaux. 2005. Temporal discounting when the choice is between two delayed rewards. *Journal of Experimental Psychology: Learning, Memory, and Cognition* 31:1121–33.

Hume, D. 1978 (1739). *Treatise of Human Nature*, ed. with an analytical index by L. A. Selby-Bigge. Oxford: Clarendon Press.

Jeffrey, R. 1965. *The Logic of Decision* (1st edn). Chicago, IL: University of Chicago Press.

Jeffrey, R. 1983. *The Logic of Decision* (2nd edn). Chicago, IL: University of Chicago Press.

Jonson, E., M. Lindorff, and L. McGuire. 2012. Paternalism and the pokies: unjustified state interference or justifiable intervention? *Journal of Business Ethics* 110:259–68.

Joyce, J. J. 1999. *Foundations of Causal Decision Theory*. Cambridge: Cambridge University Press.

Kirby, K. N., and B. Guastello. 2001. Making choices in anticipation of similar future choices can increase self-control. *Journal of Experimental Psychology: Applied* 7: 154–64.

Lewis, D. K. 1981. Causal decision theory. *Australasian Journal of Philosophy* 59:5–30. Reprinted in D. K. Lewis, *Philosophical Papers*, vol. II. Oxford: Oxford University Press; 1986: 305–39.

Nozick, R. 1969. Newcomb's problem and two principles of choice. In *Essays in Honor of Carl G. Hempel*, ed. N. Rescher (pp. 114–46). Dordrecht: Reidel.

Productivity Commission. 2010. *Gambling, Productivity Committee Enquiry Report*, vol. 1, no. 50. Canberra.

Quine, W. V. 1975. Necessary truth. In *Ways of Paradox* (pp. 68–76). Cambridge, MA: Harvard University Press.

Russell, B. 1986 (1913). On the notion of cause. *Proceedings of the Aristotelian Society* 13:1–26. Reprinted in B. Russell, *Mysticism and Logic* (pp. 173–99). London: Unwin; 1986.

Savage, L. J. 1972. *The Foundations of Statistics* (2nd edn). New York, NY: Dover.

Shafir, E., and A. Tversky. 1992. Thinking through uncertainty: nonconsequential reasoning and choice. *Cognitive Psychology* 24:449–74.

Skyrms, B. 1980. *Causal Necessity*. New Haven, CT: Yale University Press.

von Weiszäcker, C. C. 1971. Notes on endogenous change of taste. *Journal of Economic Theory* 3:345–72.

CHAPTER 5

Preference Reversals, Delay Discounting, Rational Choice, and the Brain

Leonard Green and Joel Myerson

5.1 Overview

Self-control has been viewed as the triumph of reason over impulse since the time of Plato. Contemporary philosophers frequently see preference reversals in choice between smaller, sooner rewards and larger, later ones as the paradigmatic case that needs to be explained because such reversals appear to represent failures of self-control and thus failures of reason to rein in desires for immediate gratification. Yet reversals in preference are ubiquitous. Does such common behavior illustrate a lack of rationality? After all, how can one rationally choose the larger, later reward when the times to both outcomes are long but then choose the smaller, sooner reward when the times to both are decreased equally?

Hyperbolic delay discounting functions originally were proposed as an explanation for such dynamic inconsistencies that did not require even considering the rationality of choice and as evidence against the exponential discounting functions of standard economic theory (Ainslie 1992). Contrary to the idea that preference reversals prove that discounting functions are hyperbolic, however, the discovery of robust amount effects in which larger delayed amounts are discounted less steeply than smaller ones (Green, Myerson, and McFadden 1997; Kirby 1997) raises questions concerning that argument. Importantly, preference reversals are also observed with delayed losses even though the rate at which a delayed loss is discounted is not affected by its amount (Holt et al. 2008). Moreover, nonhuman animals, which discount different amounts of reward at the same rate, also show hyperbolic discounting and preference reversals (Vanderveldt, Oliveira, and Green 2016).

If, indeed, preference reversals are the paradigmatic case that needs to be explained, then we would suggest that current findings from decision-theoretic research on this topic must be taken into account and not just the original conjectures regarding preference reversals and delay discounting.

We also would note that many common examples of preference reversals, such as the high failure rates of diets and attempts to quit smoking or the use of illicit drugs, are a special and less problematic case for rational-choice accounts because increases in the level of deprivation with the passage of time may affect the value of a sooner reward more than that of a later one. Finally, our recent findings with individuals with episodic amnesia, who show discounting similar to healthy control individuals yet are not capable of imagining the future consequences of their choices (Kwan et al. 2013), raise questions regarding the role of conscious experience in rational choice.

5.2   Introduction

The pursuit of many important long-term goals may be undermined by behaviors that lead to more immediate rewards. It is the ability to somehow avoid "succumbing to temptation," as represented by such immediate rewards, that seems central to the concept of self-control. The conflict between short-term pleasures and long-term goals has been the object of much speculation about human nature. Indeed, the notion that the capacity for reason is what sets humans apart from the "beasts" would appear to rest, at base, on the ability to determine what exactly is required to achieve long-term goals and behave accordingly. For example, Aristotle claimed that "the incontinent man, knowing that what he does is bad, does it as a result of passion, while the continent man, knowing that his appetites are bad, refuses on account of his rational principle to follow them" (Aristotle, *Nicomachean Ethics*, 1145b, 12–14, in McKeon 1941).

The ability to follow reason and not passion is, in turn, usually attributed to some force or agency (e.g., the will, or self-discipline) that controls, inspires, or holds in check the base impulses, as represented in the letters of Paul. The function of Hobbes' *Leviathan* was in large measure to rescue the governed from the vicissitudes of each other's animal, cruel, and impulsive nature. For Descartes, the need for self-control arises from conflict between some physical impulse and willpower, a spiritual force. For Freud, it was the development of the ego that opposes the id and thereby allows for delay of gratification, instituting a delay between the want and the attainment of an outcome. Current social-psychological views such as that of Muraven and Baumeister (2000) consider self-control to be some limited resource that is depleted through its use. Neuroscientific approaches posit different parts of the brain that can work together or in opposition (e.g., McClure et al. 2004). Whether the force is the spirit of God inspired by faith in

Jesus, the superego established by identification with the Oedipal father, the depletion of some limited resource within the individual, or the conflict between different parts of brain, self-control has long been, and continues to be, viewed as emanating from within the individual, the product of one part of the person holding in check the rest.

Recently, many psychologists and behavioral economists (and some philosophers) interested in self-control have focused on situations in which people must choose between a smaller, sooner reward and a larger, later one. From traditional perspectives, such as those noted earlier, the smaller, sooner reward is the temptation, and choice of the larger, later reward represents self-control. This, of course, is the basis for Mischel's famous "marshmallow test" used to study self-control in children (e.g., Mischel 1966). But adults often choose smaller, sooner rewards, too. Why? Are they just succumbing to temptation, lacking some ineffable resource?

Choosing between a smaller and a larger reward (e.g., $50 and $100) or between two rewards of the same size, one of which is available sooner (e.g., $100 now and $100 in one year), is usually straightforward, and no internal mechanism is usually invoked. When the choice is between a smaller, sooner and a larger, later reward, however, an internal mechanism then is invoked, and value judgments, which assume that the choice reflects on the character of the individual making the choice, are often posited. Choice of the smaller, sooner reward, for example, may be attributed to ignorance of the facts or to a lack of knowledge of what is best, as Socrates supposed, or to some other character flaw because, as Socrates argued, the choice "is not a matter of external behaviour, but of the inward self" (Plato, *The Republic*, Book IV, p. 141, in Cornford 1971).

Of course, in most everyday decisions, tradeoffs must be made. Some tradeoffs are like the marshmallow test situation in that they pit smaller, sooner rewards against larger, later ones. For example, do you want a car for which you could pay a lower price now or one that gets better gas mileage and saves money in the long run? Some tradeoffs, however, are not. For example, do you want the car that gets better gas mileage, or the one that has the better repair record, or the one that provides a better ride, etc., all of which concern long-term outcomes. The fact that some decisions require tradeoffs between getting a reward sooner and getting a reward later that is larger does not necessarily make those tradeoffs special. Nor do they necessarily reflect a failure within the individual who chooses the smaller, sooner reward and who may be seen as impulsive or weak willed as a result or as having a strong will or excellent self-control, if he chooses the larger, later reward. After all, the choice of a car with a better repair record or, alternatively, one with better gas mileage does not typically lead to value judgments of the kind made about individuals who choose the smaller, sooner reward.

This is not to say that all tradeoffs are equivalent. Some tradeoffs are arguably better than others, and some people may make better choices than others relative to the same goals. Nonetheless, the process of evaluating multidimensional alternatives may be the same regardless, and it is useful to have a nonjudgmental way of describing it. The concept of discounting can provide such a descriptive account of that process in the context of the tradeoff between delay and amount.

5.3   Discounting of Delayed Outcomes

In a discounting account of decision making, the value of a reward is said to decrease as the delay to its receipt increases (i.e., its value is increasingly discounted). As a consequence, a larger, later reward may have a lower (i.e., discounted) value than a smaller one if the delay to the larger one is long enough. Similarly, with aversive outcomes, a more aversive but later outcome may be discounted such that it is less negative in value than a less aversive one if, again, the delay to the more aversive outcome is long enough.

Typically, discounting is assessed by determining the present value of a delayed outcome. One way in which this is commonly done with delayed rewards is by giving an individual a series of choices between some specific amount of reward that is available after a specific delay, on the one hand, and various amounts of immediate reward, on the other hand. The point at which the individual is equally likely to choose either the immediate or the delayed reward is termed the *indifference point* and is assumed to represent the present, or subjective, value of the delayed reward (e.g., Green, Fry, and Myerson 1994; Raineri and Rachlin 1993).

Such discounting (i.e., changes in value as a function of delay) can be modeled in various ways. Before the rise of behavioral economics, economists used exponential decay functions to describe the decrease in value of a reward (or economic good) as the time until its receipt increases. In addition to its mathematical simplicity, an exponential decay function has the property, once thought desirable, of dynamic consistency. Saying that preference is dynamically consistent means that if a person prefers some amount X obtainable at time t to another amount Y at time $t + d$, then the person will show this same preference for X over Y for any value of t as long as d remains constant. This can (and has) been considered to be a fact of nature, a simplifying assumption that yields a mathematically tractable discounting function, and the consequence of the assumption that the risk that one will not receive a reward is stationary; that is, it remains constant regardless of the value of t. For purposes of comparison with alternative discounting functions, the exponential is often presented as

$$V = Ae^{-kD} \tag{5.1}$$

where V is the present value of a reward of amount A, D is the delay until its receipt, and k is a rate parameter that determines how rapidly the reward loses value as D increases.

Beginning with Mazur (1987), many psychologists, in contrast, have used hyperbolic functions to describe discounting. Such functions are usually presented either as

$$V = A/(1+kD) \tag{5.2a}$$

or as

$$V = 1/(1+kD) \tag{5.2b}$$

where, in this latter case, V represents the *relative* value of a delayed reward (expressed as a proportion of the amount of that reward). Using relative value as an index is especially helpful when comparing the discounting of delayed rewards of different amounts.

Both the exponential and hyperbolic function forms capture the fact that there is a negatively accelerated decrease in V as D increases. The hyperbolic function, however, consistently provides better fits (i.e., it accounts for more of the variance in V) than the exponential function (for a review, see Green and Myerson 2004). This may be seen in Figure 5.1, which presents a clear example showing how the exponential function often drastically underpredicts the value of a reward at long delays. This is less of a problem for the hyperbolic function, although, as may be seen, it, too, underpredicts the value of a reward at long delays.

Neither Mazur (1987) nor others provided much of a theoretical justification for the hyperbola, beyond the fact that it captures the inverse relation between value and delay and that, without the 1.0 in the denominator, the equation would predict that an immediate reward would have infinite value. Partly in response to this lack of theoretical justification and partly in response to the relatively small but systematic deviations of observed behavior from that predicted by the hyperbola, we proposed a more general function form, which we refer to as the *hyperboloid*

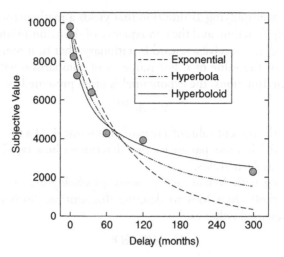

Figure 5.1 The subjective value of a delayed $10,000 reward plotted as a function of the time until its receipt. The curved lines represent alternative forms of the delay discounting function (i.e., the exponential, hyperbola, and hyperboloid; equations (5.1), (5.2a), and (5.3), respectively) fit to the data.
(*Sources:* Data are from Green, Fry, and Myerson 1994. Adapted from L. Green and J. Myerson 2004. A discounting framework for choice with delayed and probabilistic rewards. *Psychological Bulletin* 130:769–92, with permission of the American Psychological Association.)

discounting function. According to this model, the relative value of a delayed reward is given by

$$V = 1/(1+kD)^s \qquad (5.3)$$

where *s* is a parameter reflecting the nonlinear scaling of both amount and delay. The hyperboloid form is mathematically derived from three assumptions: In addition to the scaling assumption, the model assumes that Herrnstein's matching law (1970) applies and that there is always at least some minimum delay between making a choice and receiving the reward (Myerson and Green 1995).

The hyperboloid discounting function is also shown in Figure 5.1, and as may be seen, there is no longer any systematic underprediction at long delays. The hyperboloid model provides excellent fits to data from individuals and groups of different ages (e.g., Green, Fry, and Myerson 1994; Myerson and Green 1995) and from different cultures (e.g., Du, Green, and Myerson 2002), and it also more accurately describes the discounting

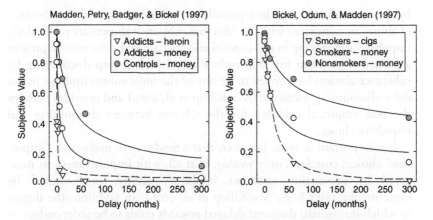

Figure 5.2 The relative subjective value of a delayed reward plotted as a function of time until its receipt. Both studies compared discounting of delayed monetary rewards by substance abusers (opioid-dependent individuals, left, and cigarette smokers, right) with discounting by control individuals. In addition, both studies compared discounting of monetary rewards with discounting of the abused substance (heroin, left, and cigarettes, right) by members of the substance-abuse group. The curved lines represent the hyperboloid discounting function (Equation 5.3) fit to the data. (*Sources:* Data are from Madden et al. 1997 and Bickel, Odum, and Madden 1999. Reprinted from L. Green and J. Myerson. 2004. A discounting framework for choice with delayed and probabilistic rewards. *Psychological Bulletin* 130:769–92, with permission of the American Psychological Association.)

of both real and hypothetical rewards, as well as different types of rewards (e.g., vacations, free use of a car, immediately consumable commodities; Estle et al. 2007; Jimura et al. 2009; Raineri and Rachlin, 1993) than the simple hyperbola, which is the special case of the hyperboloid when $s = 1.0$ (i.e., scaling is linear; for a review, see Green and Myerson 2004). Two particularly relevant examples are given in Figure 5.2, which shows the discounting of delayed heroin (Madden et al. 1997) and cigarette (Bickel, Odum, and Madden 1999) rewards by addicts and smokers, respectively, and also compares the discounting of delayed monetary rewards by the substance abusers and control individuals.

As may be seen in Figure 5.2, the groups of substance abusers showed much steeper discounting of monetary rewards than the control groups. Also relevant is the fact that when the delayed reward was some amount of money or an amount of their substance of abuse of equivalent monetary value, the substance abusers discounted the abused substance much more steeply. Both these findings, the steeper discounting of monetary rewards by substance abusers and the especially steep discounting of their substance of abuse, are consistent with the idea that substance abusers are particularly impulsive and lacking in self-control and that abused substances represent a particular problem for such individuals. Indeed, steep discounting by substance abusers has proven to be one of the most robust findings in the delay-discounting literature (MacKillop et al. 2010) and provides perhaps the best empirical evidence for the relation between discounting and impulsive choice.

It is important to note, however, that a tendency to make such "impulsive" choices correlates only weakly, if at all, with impulsiveness, as measured by personality tests, or inhibitory ability, as measured by experimental tasks (e.g., MacKillop et al. 2016). In addition, the degree to which individuals discount delayed rewards tends to be independent of their tendency to discount delayed losses (i.e., payments). Moreover, given that the best evidence for the relation between discounting and impulsive choice comes from the finding that substance abusers are steeper discounters of delayed rewards than control individuals, it is notable that substance abusers appear not to differ from control individuals in their discounting of delayed losses. This may be seen in Figure 5.3, which shows data from a recent study with alcohol-dependent, low-income African Americans (Myerson et al. 2015). A similar pattern of results was obtained in a study with cocaine-dependent Mexican men (i.e., steeper discounting of delayed rewards but no difference in degree of discounting between the substance abusers and control individuals in their discounting of delayed losses; Mejía-Cruz et al. 2016).

The substance abusers also did not differ from the control individuals in the discounting of probabilistic rewards (see bottom panel of Figure 5.3). Moreover, the discounting of delayed rewards is not negatively correlated with the discounting of probabilistic rewards, as might be expected if risk-taking was another aspect of impulsive choice (for a review, see Green and Myerson 2010). Although the discounting of some types of delayed rewards is correlated (for a review, see Odum 2011), there are notable exceptions, for example, the very weak correlations observed between the discounting of delayed hypothetical monetary rewards and delayed hypothetical health outcomes (Chapman 1996) and the lack of correlation between the discounting of delayed hypothetical monetary rewards and delayed real liquid rewards (Jimura et al. 2011).

Finally, it should be noted that the evidence from the steep discounting of delayed rewards by substance abusers is correlational and thus leaves open the possibility that the relation between substance abuse and

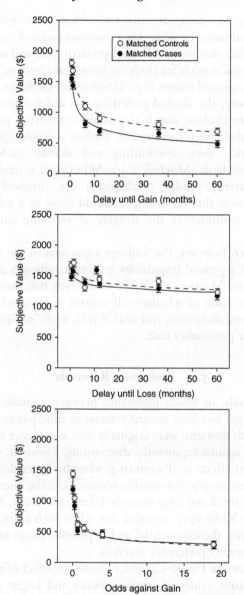

Figure 5.3 Discounting of delayed gains (top), delayed losses (center), and probabilistic gains (bottom) by alcohol-dependent African Americans and control individuals matched on age, gender, and education. Curved lines represent the hyperboloid discounting function (Equation 5.3) fit to the data. Error bars represent the standard error of the mean.

(*Source:* Reprinted from J. Myerson, L. Green, C. van den Berk-Clark, and R. A. Grucza. 2015. Male, but not female, alcohol-dependent African Americans discount delayed gains more steeply than propensity-score matched controls. *Psychopharmacology* 232:4493-4503, with permission of Springer.)

discounting is due to a common cause rather than short-sighted decisions by the substance abusers. Evidence for a common cause of a genetic nature comes from the fact that various alcohol-preferring rat and mouse strains, often used as animal models for studying human alcoholism, show steeper discounting than control strains (e.g., Oberlin and Grahame 2009; Perkel et al. 2015). Clearly, the alcohol-preferring rats did not get that way by making impulsive choices, nor is the alcohol-preferring phenotype the consequence of their steep discounting. Rather, it would appear that the two characteristics, steep discounting and alcohol preference, have a common genetic basis (Mitchell 2011). Although it is unclear whether, or to what extent, similar mechanisms are involved in human alcohol dependence, this rat example should serve as a cautionary tale and a relevant reminder of the dangers of inferring causation from correlation.

Taken together, however, the findings argue against the idea that discounting reflects a general impulsivity trait that indicates a lack of self-control (Green and Myerson 2013). We would note that these findings do not speak to the issue of whether self-control is involved in decisions concerning delayed outcomes, just that if it is, such "self-control" is not a general, unitary personality trait.

5.4   Preference Reversals

Preference reversals, in which preference between a smaller but sooner reward and a larger but later reward reverses as time passes and the wait until both rewards shortens, were originally seen as evidence for hyperbolic discounting and against exponential discounting. However, the discovery of robust amount effects in discounting, whereby larger delayed rewards are discounted less steeply than smaller rewards (i.e., the rates at which the relative values of small and large rewards differ; e.g., Green, Myerson, and McFadden 1997; Kirby 1997) revealed that when such amount effects are taken into account, the exponential and hyperbolic forms of discounting functions both predict preference reversals.

This may be seen in Figure 5.4, which shows the effect of the passage of time on the relative values of smaller, sooner and larger, later rewards predicted by a hyperbolic discounting function (top) and the exponential discounting function under the (incorrect) assumption that smaller and larger rewards are discounted at the same rate (center) and under the

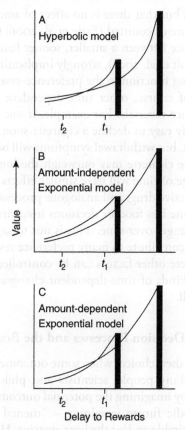

Figure 5.4 Subjective value of a reward as a function of delay to its receipt. The heights of the bars represent the amounts of the two rewards, and the curved lines represent their value at different points in time. Crossovers indicate points of preference reversal. Separate panels show the predictions of hyperbolic (A), amount-independent exponential (B), and amount-dependent exponential (C) discounting models. Note that A and C both depict preference reversals because the value of the large reward exceeds that of the small reward when the delay to both rewards is relatively long (e.g., at t_2), but the value of the small reward exceeds that of the large reward when the delay to both rewards is relatively brief (e.g., at t_1).
(*Source:* Reprinted from L. Green and J. Myerson. 1996. Exponential versus hyperbolic discounting of delayed outcomes: risk and waiting time. *American Zoologist* 36:496–505, by permission of Oxford University Press.)

(correct) assumption that smaller rewards are discounted at steeper rates (bottom). More recently, however, we have demonstrated that the discounting of delayed losses is also well described by hyperboloid discounting functions but that there is no effect of amount on the rate at which delayed losses are discounted (Estle et al. 2006; Green et al. 2014). Nevertheless, preference between a smaller, sooner loss and a larger, later loss often reverses (Holt et al. 2008), strongly implicating the hyperboloid form of the discounting function in the preference-reversal phenomenon.

In everyday life, of course, other time-dependent changes also may contribute to preference reversals. For example, if one decides to give up smoking, it is relatively easy to decline a cigarette soon after having made the decision to abstain, but withdrawal symptoms will begin later, until the value of an immediate cigarette may outweigh the long-term benefits of refraining. The passage of time will have similar effects on the decision to abstain from any addictive drug, and analogous processes may be involved in decisions to consume less food or decisions involving any type of outcome whose value changes over time. This is not to say that hyperboloid discounting does not contribute to many preference reversals, particularly in the laboratory, where other factors can be controlled, but that outside the laboratory other kinds of time-dependent changes in outcome value may contribute as well.

5.5   Decision Processes and the Brain

How do people make their choices when some outcomes occur sooner and others occur later? Many people, scientists and philosophers included, believe that we do so by imagining the potential outcomes of our decisions (i.e., engage in episodic future imagining – "mental time travel") and choosing the most desirable and/or the least aversive. However, our recent findings with individuals with episodic amnesia, who show discounting similar to healthy control individuals yet are not capable of imagining the future consequences of their choices (Kwan et al. 2013), raise fundamental questions regarding the role of conscious experience in choice – rational or not.

In a series of studies conducted in collaboration with Shayna Rosenbaum, Carl Craver, and Donna Kwan, we have taken advantage of the specific characteristics of individuals with episodic amnesia as a result of brain injury (or in some cases, developmental differences) to examine the contribution of future imagining to decision making. Our first study (Kwan et al. 2012) focused on K.C., a well-known and extensively documented amnesic individual who sustained bilateral hippocampal damage in a 1981 motorcycle accident and as a result, was unable to recall any past personal event. Importantly, he also was unable to imagine future personal events (Rosenbaum et al. 2005). K.C.'s injury left him with a unique neuropsychological profile. For example, when cued with respect to five upcoming events in his life, he was not able to imagine details of what would happen for any one of them. K.C.'s performance represents a striking deficit and differs considerably from what is observed even in patients with probable Alzheimer's disease (Addis et al. 2009).

Interestingly, despite such severe impairment, K.C. remembered facts about himself and the world and functioned well in many cognitive domains (Rosenbaum et al. 2007). When tested after his injury, K.C. had an average IQ and relatively preserved cognitive functioning outside of his episodic memory impairment and deficit in future imagining (Rosenbaum et al. 2005, 2009). Thus, he was a perfect test case for examining the role that imagining the future plays in decision making.

It is typically assumed that the ability to imagine one's personal future is a fundamental aspect of future-oriented decision making and therefore that performance on discounting tasks reflects this ability. Thus, it is of considerable interest how someone who has extensive brain damage that left him unable to construct details of personal future events might perform on a task that requires valuation of future rewards. Although from the usual perspective one might not expect K.C., or anyone with a similar deficit in future imagining, to show systematic discounting, it is not necessarily obvious what he should do instead. For example, it might be predicted that if someone cannot imagine receiving a future reward, then the rational choice would be to always choose an immediate reward over a delayed one regardless of the size of the delayed rewards (see, e.g., Boyer 2008). Alternatively, it might be argued that people imagine the wait period itself and that the anticipated unpleasantness of waiting for a delayed reward biases subjects toward immediate rewards (Luhmann 2009). According to this view, K.C., unlike control individuals, should not show a bias for a smaller, immediate reward and instead should always choose the larger amount. In either case, there should be no systematic effect of delay on K.C.'s choices.

For our first study (Kwan et al. 2012), we had K.C. and 18 healthy age- and education-matched control individuals complete a standard discounting task, on which they made a series of choices between hypothetical monetary offers – a smaller, immediate amount and a larger, future amount. Participants were tested on multiple occasions to assess the consistency of K.C.'s performance relative to the control individuals. Contrary to both preceding hypotheses, the subjective value that K.C. placed on a future reward decreased systematically with the delay to the receipt of the reward. Moreover, the degree to which he discounted such rewards, as assessed using the area-under-the-curve measure, was close to the median of the control group for both the $100 and $2,000 delayed rewards tested. Notably, K.C., like the control individuals, showed a magnitude effect (i.e., shallower discounting of the larger delayed amount), a standard finding in the human delay discounting literature (Green and Myerson 2004).

These results demonstrate that K.C. still valued future rewards despite being unable to construct the details of either past or future events. This finding is inconsistent with the predictions of accounts that emphasize the critical role of imagining future events in making choices when the decisions involve future outcomes (Boyer 2008; Luhmann et al. 2008).

Interestingly, in another study (Craver et al. 2014), K.C. showed normal performance (i.e., choice behavior indistinguishable from that of healthy control individuals) on tasks involving choices between probabilistic gains (hypothetical monetary rewards) and between probabilistic losses (payments). These tasks were of particular interest because they also have been assumed to involve future-oriented decision making. The capacity to anticipate future experiences of regret has been hypothesized to explain otherwise irrational aspects of human decision making, including the certainty effect (Kahneman and Tversky 1979) and the common-ratio effect (Allais 1953). For example, people regularly choose a certain reward over a risky option that has greater expected utility (e.g., choose $3,000 for certain over $4,000 with a probability of 0.80), a finding termed the *certainty effect* (Kahneman and Tversky 1979). According to Loomes and Sugden (1982), this occurs because people anticipate the feelings of regret they would experience if they chose the risky option and it did not pay off.

Loomes and Sugden (1982) also argued that anticipated regret theory explains the common-ratio effect first identified by Allais (1953). This effect also is irrational, viewed from the perspective of normative economic theory, because it violates the substitution axiom, one of the fundamental principles of expected utility theory (MacCrimmon and Larsson 1979). Loomes and Sugden showed mathematically that if one makes certain assumptions about anticipation of regret, then expected utility theory can accommodate both the certainty effect and the common-ratio effect.

Based on anticipated regret theory, one might expect that, paradoxically, an individual such as K.C., who is incapable of episodically imagining his personal future and thus does not anticipate feeling regret, would show more rational decision making.

As presented by Kahneman and Tversky (1979), the common-ratio effect is exemplified by the fact that when offered a choice between $3,000 for sure and an 80 percent chance of winning $4,000, the majority of people choose the sure thing (as in the certainty effect), but when the probabilities are reduced by a common factor (e.g., a 25 percent chance of winning $3,000 versus a 20 percent chance of winning $4,000), the majority now pick the riskier gamble (i.e., the one with the lower probability of winning). Such choices represent a clear violation of expected utility because multiplying the probabilities of the outcomes by a constant does not change their relative expected utilities, yet preferences reverse. Loomes and Sugden (1982) explicitly argued for their theory because it continues to treat individuals as rational despite what would appear to be violations of classical expected utility. At stake for them was the very foundation of classical economic theory, the assumption that *Homo economicus* is a rational decision maker. Loomes and Sugden call on the anticipation of regret to turn apparently irrational decisions rational. Indeed, Loomes and Sugden argued that a person who did not anticipate regret would show neither the certainty effect nor the common-ratio effect.

To test this view, we again studied K.C., who was profoundly deficient in his ability to generate future and fictitious scenarios, including those involving regret. K.C. and 12 control participants were presented with a series of choices between two hypothetical investments. For example, one trial involved a choice between a 90 percent chance of earning $15,000 (with a 10 percent chance of earning nothing) versus a 100 percent chance of earning $10,000. Another trial, matched on the ratio of the expected values of the choice alternatives, involved a 9 percent chance of earning $15,000 (with a 91 percent chance of earning nothing) versus a 10 percent chance of earning $10,000 (with a 90 percent chance of earning nothing). Choice of the certain gain, despite its lower expected value, on the first trial represents the certainty effect. Allais (1953) showed that when presented with choices such as those in the first and second trials, healthy individuals show a preference change (the common-ratio effect) that, according to normative economic theory, is irrational because the ratio of the expected values of the choice alternatives, 1.35 to 1.00, is the same in both trials.

Contrary to what Loomes and Sugden (1982) would predict based on their regret hypothesis, K.C., who had nearly complete deficits in his

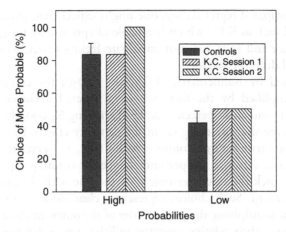

Figure 5.5 Percent choice of the more probable reward by K.C. on each of two testing sessions and by control participants. Error bars represent the standard error of the mean.
(*Source:* Reprinted from C. Craver, F. Cova, L. Green, et al. 2014. An Allais paradox without mental time travel. *Hippocampus* 24:1375–80, with permission of John Wiley and Sons.)

ability to imagine his personal future, nonetheless displayed the certainty effect, choosing the certain alternative on choices such as those in the first trial despite its lower expected value (see Figure 5.5, left). Moreover, he also showed the common-ratio effect on pairs of trials such as those in the example given earlier, switching his preference even though the two options on such trials have the same expected value (Figure 5.5, right). The point here is not that K.C. showed "irrational" behavior; rather, as may be seen in Figure 5.5, K.C.'s behavior was indistinguishable from that of healthy control individuals. These results suggest that the episodic anticipation of future regret does not generally explain the certainty and common-ratio effects, which are both considered irrational from the perspective of normative economic theory, and raise the question of the role of anticipation in future-oriented decisions more generally.

We would note that K.C.'s choice behavior is not unique. We have shown that other brain-damaged individuals who have profound deficits in their ability to imagine future personal events exhibit choice behavior similar to his. Importantly, this behavior does not differ in significant ways from that of demographically matched healthy individuals, strongly suggesting that future-oriented decision making need not depend on the ability to imagine future events. In another collaborative study with

Shayna Rosenbaum, we tested K.C. and three other amnesic individuals with deficits in future imagining on both a delay discounting task to assess their valuation of future rewards and a probability discounting task to assess their risk-taking (Kwan et al. 2013). It is to be noted that two of the additional amnesic individuals (D.A. and D.G.), like K.C., had adult-onset episodic amnesia, and the other individual (H.C.) had early-onset amnesia and never developed normal episodic memory.

Figure 5.6 presents the findings from K.C. and D.A. and the median values for the respective control individuals [see Kwan et al. (2013) for the results for the other amnesic individuals and their respective controls]. As may be seen in the left-hand panels, the amnesic individuals (like the controls) exhibited clear discounting of both a $100 and $2,000 future reward. In each case, the subjective value that the amnesic individuals placed on the reward decreased systematically as the delay until receiving the reward increased. Moreover, no significant differences were observed between any of the amnesic individuals and their respective controls.

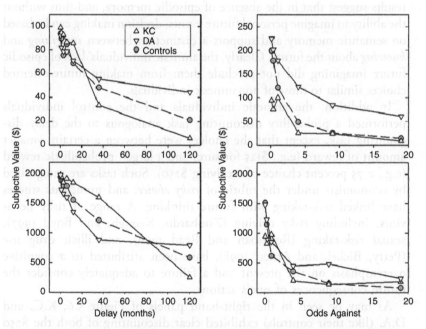

Figure 5.6 Subjective value as a function of delay (left) and odds against (right) for K.C., D.A., and their controls. The top row depicts the discounting of the smaller ($100 delayed and $250 probabilistic) reward, and the bottom row depicts the discounting of the larger ($2,000) reward (see Kwan et al. 2013).

So how did K.C. and the other amnesic individuals make their decisions on the discounting task? In an interview after the discounting task, K.C. was unable to imagine episodes or ways in which he might spend the hypothetical payout; when probed, K.C. consistently stated that he supposed he would "put it in the bank" and reported a "blank" state of mind when asked to construct ways in which he might use future rewards that he had chosen over immediate rewards. When asked about his overall strategy, he reported relying on a gut feeling to choose "the best deal."

As for the other three amnesic individuals tested, H.C. described her strategy as simply determining "how long [she] would be able to hold out." D.A. reported using a strictly economic strategy, specifically estimating inflation and interest rates, but did not expand on how he made his calculations. When probed, D.A. gave vague and general examples of how he might spend the money, such as "going on a vacation." D.G. reported "wondering if [he] could wait that long" when deciding between immediate and delayed rewards, and when pressed further, he reported that his decisions were based on a "gut feeling." Taken together, these results suggest that in the absence of episodic memory, and thus without the ability to imagine personal future events, decision making can be based on semantic memory and support a distinction between *imagining* and *knowing* about the future. Clearly, the amnesic individuals' lack of episodic future imagining did not preclude them from making future-oriented choices similar to those of nonamnesic individuals.

In addition, the amnesic individuals and the control individuals performed a probability discounting task analogous to the delay discounting task, except that the choices were between a certain, smaller amount of reward (e.g., $125 for sure) and a larger, probabilistic reward (e.g., a 75 percent chance of receiving $250). Such tasks are categorized by economists under the rubric of *risky choice*, and numerous studies have linked risk-taking with future thinking. A range of risky behaviors, including risky driving (Zimbardo, Keough, and Boyd 1997), sexual risk-taking (Rothspan and Read 1996), and illicit drug use (Petry, Bickel, and Arnett 1998), have been attributed to a cognitive overemphasis on the present and a failure to adequately consider the future consequences of one's actions.

As may be seen in the right-hand panels of Figure 5.6, K.C. and D.A. (like their controls) exhibited clear discounting of both the $250 and $2,000 probabilistic rewards. In each case, the subjective value that the amnesic individuals placed on a reward decreased systematically as its probability decreased and, accordingly, the odds against its receipt increased. There were no significant differences in probability discounting between the amnesic individuals and their respective control individuals. These results indicate that despite their inability to imagine their future experiences, the amnesic individuals can make decisions involving probabilistic outcomes similar to those of healthy control individuals, just as they can make apparently normal decisions about future outcomes on delay discounting tasks. This is not to say that their behavior was always rational; rather, their behavior was no more rational or irrational than those of healthy control individuals.

K.C., D.A., H.C., and D.G. also were administered the Zimbardo Time Perspective Inventory (ZTPI; Zimbardo and Boyd 1999) to assess their personal orientation toward the past, present, and future. Biases in time perspective have been hypothesized to influence one's judgments, decisions, and actions. A present orientation as measured by the ZTPI, for example, has been reported to be predictive of risky driving (Zimbardo et al. 1997) and the use of substances such as alcohol, drugs, and tobacco (Keough, Zimbardo, and Boyd 1999). If the ability to construct future scenes and scenarios and imagine future experiences is a prerequisite for a future-oriented time perspective, then one might expect that individuals with amnesia will adopt a present-centered, hedonistic attitude that minimizes consideration of future events. None of the amnesic individuals, however, identified with the present-biased perspectives reflected in the ZTPI, and indeed, none was more strongly biased toward the present than the control individuals.

Taken together, these findings provide insight into the strategies and decision-making processes underlying widely used tasks such as delay and probability discounting that are often interpreted as measuring impulsivity and self-control. The fact that we find no difference in delay or probability discounting between amnesic individuals and healthy control individuals indicates that decision making on such tasks does not require an ability to reexperience or construct any specific past or future event in memory or imagination, although, of course, it does not preclude the possibility that individuals make their decisions in this way at least some of the time.

5.6 Discounting, Self-Control, and Rational Choice

As mentioned earlier, psychologists frequently describe choosing a larger, later reward over a smaller one available sooner as reflecting self-control, in contrast to choosing the smaller, sooner reward, which is described as impulsive. A more nonjudgmental approach to such behavior, however, is to characterize choices of smaller, sooner rewards over larger, later ones as reflecting the discounting of the value of delayed outcomes, where 'discounting' refers to the decrease in the subjective value of an outcome as the time until its occurrence increases.

As Bermúdez (Chapter 8) has noted more generally, decision-theoretic approaches (and the discounting framework on which the present effort is based clearly belongs in that category) implicitly challenge usual conceptions of self-control. After all, if an individual is always choosing the alternative that has the higher value at that moment, then how can that choice represent self-control, even when the individual chooses a larger, delayed reward over a smaller, sooner one? In contrast, if the individual choosing the larger, later reward does not value it more highly at the moment of choice, then how can that be rational?

In addition to this philosophical conundrum, the notion of self-control as some effortful process involving exercise of willpower is not holding up empirically. Researchers looking for subjective correlates of the exertion of self-control by querying participants during their daily activities have found that, contrary to expectation, individuals who appear to succumb to temptation the least report exerting the least effort doing so, apparently due to their use of better strategies, not greater willpower (Hofman et al. 2012), even in the case of individuals recovering from addiction (Snoek, Levy, and Kennett 2016).

Moreover, the notion of self-control tends to lead psychologists, at least, to describe the choice of the alternative that they typically frown on (the smaller, sooner reward) in pejorative terms (Green and Myerson 1993). For example, choice of a smaller, sooner reward is described as an *impulsive choice*. Further, this pejorative language is then often extended to the individual who is considered impulsive, short-sighted, and lacking in willpower. Again, this view fails on empirical grounds. As already noted, factor analyses reveal that the tendency to steeply discount delayed rewards is unrelated to performance on standardized tests of impulsiveness and is also unrelated to the ability to inhibit prepotent responses (e.g., MacKillop et al. 2016).

Finally, as William S. Burroughs, writing as William Lee (1953), noted in his novel *Junkie: Confessions of an Unredeemed Drug Addict*, "A junkie spends half his life waiting." Although drug addicts sometimes have drugs immediately available, the recurrent need to purchase more drugs is time-consuming and frequently requires long waits for drugs as well as much planning. Indeed, monkeys addicted to cocaine and given a choice between immediate food and delayed cocaine show very shallow discounting, even though when given a choice between immediate cocaine and delayed food, they show very steep discounting (Huskinson et al. 2016, 2015). The latter choice situation models that faced by human addicts when their drugs are immediately available, and the monkeys' behavior might be labeled impulsive, whereas the former situation models the process of obtaining drugs, and here the monkeys' behavior could be described as showing great self-control. Such research would probably not be allowed if the subjects were humans, but the use of monkeys presumably reveals the fundamental processes involved in such behavior, which is not easily categorized either as impulsive or as reflecting self-control.

In defense of the impulsive/self-control characterization, it has been reported that the degree of discounting predicts various problem behaviors, ranging from poor academic performance (e.g., Kirby, Winston, and Santiesteban 2005) to eating disorders (e.g., Manwaring et al. 2011), in addition to substance abuse, and choice of the smaller reward under certain circumstances might be termed *irrational*, although a better term might be *suboptimal*. It is not irrational or even suboptimal, however, under other circumstances. For example, if you have a one-time opportunity to choose between $200 now and $1,000 in a year, choosing more, even though you would have to wait for it, may make sense. But if you must get a part replaced on your car in order to get to work, and the total cost of the repair would be $200 that you do not have any other means of obtaining, then choosing the smaller immediate amount may be more than rational – it may be absolutely necessary.

Such situations are well studied in behavioral biology, where the iconic situation is that of a small bird in winter. Birds have high metabolic rates and need constant food to survive, particularly in cold weather, and cannot always "afford" to choose the risky alternative with the higher expected value or to delay food consumption so as to obtain more food later. Similar situations, and not just one's degree of *impulsivity*, underlie the payday loan industry. In fact, Shah, Mullainathan, and Shafir (2012) argue that resource scarcity changes how people look at problems and make decisions. When money is abundant, for example, basic expenses are handled easily as they arise, but when it is scarce, such expenses create immediate problems that necessarily lead to the neglect of other problems, including those concerned with meeting long-term goals.

Shah et al. (2012) stressed that this view is not specific to poverty, nor does it make assumptions about the personality characteristics of the poor (e.g., that they are impulsive). Rather, it deals with the problem of scarcity more generally, including, for example, scarcity of time. Whereas the poor may borrow money to meet immediate expenses, the busy may borrow time from other tasks because they need to meet immediate goals. Importantly, this approach, like ours, does not involve judgments about the personal characteristics of the individuals involved, beyond the fact that they are poor, or busy, or are experiencing scarcity of some other critical resource. In a series of studies involving hypothetical budgets, Shah et al. showed that individuals randomly assigned tight budgets made decisions indicating that they were focused on commodities that were scarce, to the exclusion of other goods and benefits. So, too, the small bird in winter, the junkie, and likely most of us, at least some of the time, make choices that others might consider impulsive, irrational, and reflecting a lack of self-control, without considering the larger "economic" context in which those decisions are being made.

References

Addis, D. R., D. C. Sacchetti, B. A. Ally, A. E. Budson, and D. L. Schacter. 2009. Episodic simulation of future events is impaired in mild Alzheimer's disease. *Neuropsychologia* 47:2660–71.

Ainslie, G. 1992. *Picoeconomics: The strategic interaction of successive motivational states within the person.* Cambridge: Cambridge University Press.

Allais, M. 1953. Le comportement de l'homme rationnel devant le risque: critique des postulats et axiomes de l'Ecole Americaine. *Econometrica* 21:503–46.

Bermúdez, J. L. 2018. Frames, rationality, and self-control. In *Self-Control, Decision Theory, and Rationality*, ed. J. L. Bermúdez (pp. 179–203). Cambridge: Cambridge University Press.

Bickel, W. K., A. L. Odum, and G. J. Madden. 1999. Impulsivity and cigarette smoking: delay discounting in current, never, and ex-smokers. *Psychopharmacology* 146:447–54.

Boyer, P. 2008. Evolutionary economics of mental time travel? *Trends in Cognitive Sciences* 12:219–24.

Chapman, G. B. 1996. Temporal discounting and utility for health and money. *Journal of Experimental Psychology: Learning, Memory, and Cognition* 22:771–91.

Cornford, F. M., trans. 1971. *The Republic of Plato.* London: Oxford University Press.

Craver, C., F. Cova, L. Green, et al. 2014. An Allais paradox without mental time travel. *Hippocampus* 24:1375–80.

Du, W., L. Green, and J. Myerson. 2002. Cross-cultural comparisons of discounting delayed and probabilistic rewards. *Psychological Record* 52:479–92.

Estle, S. J., L. Green, J. Myerson, and D. D. Holt, 2006. Differential effects of amount on temporal and probability discounting of gains and losses. *Memory & Cognition* 34:914–28.

Estle, S. J., L. Green, J. Myerson, and D. D. Holt. 2007. Discounting of monetary and directly consumable rewards. *Psychological Science* 18:58–63.

Green, L., A. Fry, and J. Myerson. 1994. Discounting of delayed rewards: a life-span comparison. *Psychological Science* 5:33.

Green, L., and J. Myerson. 1993. Alternative frameworks for the analysis of self control. *Behavior and Philosophy* 21:37–47.

Green, L., and J. Myerson. 1996. Exponential versus hyperbolic discounting of delayed outcomes: risk and waiting time. *American Zoologist* 36:496–505.

Green, L., and J. Myerson. 2004. A discounting framework for choice with delayed and probabilistic rewards. *Psychological Bulletin* 130:769–92.

Green, L., and J. Myerson. 2010. Experimental and correlational analyses of delay and probability discounting. In *Impulsivity: The Behavioral and Neurological Science of Discounting*, ed. G. J. Madden and W. K. Bickel (pp. 67–92). Washington, DC: American Psychological Association.

Green, L., and J. Myerson. 2013. How many impulsivities? A discounting perspective. *Journal of the Experimental Analysis of Behavior* 99:3–13.

Green, L., A. F. Fry, and J. Myerson. 1994. Discounting of delayed rewards: a life-span comparison. *Psychological Science* 5:33–36.

Green, L., J. Myerson, and E. McFadden. 1997. Rate of temporal discounting decreases with amount of reward. *Memory & Cognition* 25:715–23.

Green, L., J. Myerson, L. Oliveira, and S. E. Chang. 2014. Discounting of delayed and probabilistic losses over a wide range of amounts. *Journal of the Experimental Analysis of Behavior* 101:186–200.

Herrnstein, R. J. 1970. On the law of effect. *Journal of the Experimental Analysis of Behavior* 13:243–66.

Hofmann, W., R. F. Baumeister, G. Förster, and K, D. Vohs. 2012. Everyday temptations: an experience sampling study of desire, conflict, and self-control. *Journal of Personality and Social Psychology* 102L:1318–35.

Holt, D. D., L. Green, J. Myerson, and S. J. Estle. 2008. Preference reversals with losses. *Psychonomic Bulletin & Review* 15:89–95.

Huskinson, S. L., J. Myerson, L. Green, et al. 2016. Shallow discounting of delayed cocaine by male rhesus monkeys when immediate food is the choice alternative. *Experimental and Clinical Psychopharmacology* 24:456–63.

Huskinson, S. L., W. L. Woolverton, L. Green, J. Myerson, and K. B. Freeman. 2015. Delay discounting of food by rhesus monkeys: cocaine and food choice in isomorphic and allomorphic situations. *Experimental and Clinical Psychopharmacology* 23:184–93.

Jimura, K., J. Myerson, J. Hilgard, T. S. Braver, and L. Green. 2009. Are people really more patient than other animals? Evidence from human discounting of real liquid rewards. *Psychonomic Bulletin & Review* 16:1071–75.

Jimura, K., J. Myerson, J. Hilgard, et al. 2011. Domain independence and stability in young and older adults' discounting of delayed rewards. *Behavioural Processes* 87:253–59.

Kahneman, D., and A. Tversky. 1979. Prospect theory: an analysis of decision under risk. *Econometrica* 47:263–91.

Keough, K. A., P. G. Zimbardo, and J. N. Boyd. 1999. Who's smoking, drinking, and using drugs? Time perspective as a predictor of substance use. *Basic and Applied Social Psychology* 21:149–64.

Kirby, K. N. 1997. Bidding on the future: evidence against normative discounting of delayed rewards. *Journal of Experimental Psychology: General* 126:54–70.

Kirby, K. N., G. C. Winston, and M. Santiesteban. 2005. Impatience and grades: delay-discount rates correlate negatively with college GPA. *Learning and Individual Differences* 15:213–22.

Kwan, D., C. F. Craver, L. Green, J. Myerson, and R. S. Rosenbaum. 2013. Dissociations in future thinking following hippocampal damage: evidence from discounting and time perspective in episodic amnesia. *Journal of Experimental Psychology: General* 142:1355–69.

Kwan, D., C. F. Craver, L. Green, et al. 2012. Future decision-making without episodic mental time travel. *Hippocampus* 22:1215–19.

Lee, W. 1953. *Junkie: Confessions of an Unredeemed Drug Addict.* New York, NY: Ace Books.

Loomes, G., and R. Sugden. 1982. Regret theory: an alternative theory of rational choice under uncertainty. *Economic Journal* 92:805–24.

Luhmann, C. C. 2009. Temporal decision-making: insights from cognitive neuroscience. *Frontiers in Behavioral Neuroscience* 3:1–9.

Luhmann, C. C., M. Chun, D. Y. Yi, D. Lee, and X. J. Wang. 2008. Neural dissociation between delay and uncertainty intertemporal choice. *Journal of Neuroscience* 28:14459–66.

MacCrimmon, K. R., and S. Larsson. 1979. Utility theory: axioms versus "paradoxes." In *Expected Utility and the Allais Paradox*, ed. M. Allais and O. Hagen (pp. 333–409). Boston, MA: Reidel.

MacKillop, J., R. Miranda Jr, P. M. Monti, et al. 2010. Alcohol demand, delayed reward discounting, and craving in relation to drinking and alcohol use disorders. *Journal of Abnormal Psychology* 119:106–14.

MacKillop, J., J. Weafer, J. C. Gray, et al. 2016. The latent structure of impulsivity: impulsive choice, impulsive action, and impulsive personality traits. *Psychopharmacology* 233:3361–70.

Madden, G. J., N. M. Petry, G. J., Badger, and W. K. Bickel. 1997. Impulsive and self-control choices in opioid-dependent patients and non-drug-using control participants: drug and monetary rewards. *Experimental and Clinical Psychopharmacology* 5:256–62.

Manwaring, J. L., L. Green, J., Myerson, M. J. Strube, and D. E. Wilfley. 2011. Discounting of various types of rewards by women with and without binge eating disorder: evidence for general rather than specific differences. *Psychological Record* 61:561–82.

Mazur, J. E. 1987. An adjusting procedure for studying delayed reinforcement. In *Quantitative Analyses of Behavior*, vol. 5: *The Effect of Delay and of Intervening Events on Reinforcement Value*, ed. M. L. Commons, J. E. Mazur, J. A. Nevin, and H. Rachlin (pp. 55–73). Hillsdale, NJ: Erlbaum.

McClure, S. M., D. I. Laibson, G. Loewenstein, and J. D. Cohen. 2004. Separate neural systems value immediate and delayed monetary rewards. *Science* 306: 503–7.

McKeon, R., ed. 1941. *The Basic Works of Aristotle*. New York, NY: Random House.

Mejía-Cruz, D., L. Green, J. Myerson, S. Morales-Chainé, and J. Nieto. 2016. Delay and probability discounting by drug-dependent cocaine and marijuana users. *Psychopharmacology* 233:2705–14.

Mischel, W. 1966. Theory and research on the antecedents of self-imposed delay of reward. In *Progress in Experimental Personality Research*, ed. B. A. Maher, vol. 3 (pp. 85–132). New York, NY: Academic Press.

Mitchell, S. H. 2011. The genetic basis of delay discounting and its genetic relationship to alcohol dependence. *Behavioural Processes* 87:10–17.

Muraven, M., and R. F. Baumeister. 2000. Self-regulation and depletion of limited resources: does self-control resemble a muscle? *Psychological Bulletin* 126: 247–59.

Myerson, J., and L. Green. 1995. Discounting of delayed rewards: models of individual choice. *Journal of the Experimental Analysis of Behavior* 64: 263–76.

Myerson, J., L. Green, C. van den Berk-Clark, and R. A. Grucza. 2015. Male, but not female, alcohol-dependent African Americans discount delayed gains more steeply than propensity-score matched controls. *Psychopharmacology* 232: 4493–503.

Oberlin, B. G., and N. J. Grahame. 2009. High-alcohol preferring mice are more impulsive than low-alcohol preferring mice as measured in the delay discounting task. *Alcoholism: Clinical and Experimental Research* 33:1294–303.

Odum, A. L. 2011. Delay discounting: I'm a k, you're a k. *Journal of the Experimental Analysis of Behavior* 96:427–439.

Perkel, J. K., B. S. Bentzley, M. E. Andrzejewski, and M. P. Martinetti. 2015. Delay discounting for sucrose in alcohol-preferring and nonpreferring rats using a sipper tube within-sessions task. *Alcoholism: Clinical and Experimental Research* 39:232–38.

Petry, N. M., W. K. Bickel, and M. Arnett. 1998. Shortened time horizons and insensitivity to future consequences in heroin addicts. *Addiction* 93:729–38.

Raineri, A., and H. Rachlin. 1993. The effect of temporal constraints on the value of money and other commodities. *Journal of Behavioral Decision Making* 94:77–94.

Rosenbaum, R. S., A. Gilboa, B. Levine, G. Winocur, and M. Moscovitch. 2009. Amnesia as an impairment of detail generation and binding: evidence from personal, fictional, and semantic narratives in KC. *Neuropsychologia* 47: 2181–87.

Rosenbaum, R. S., S. Köhler, D. L. Schacter, et al. 2005. The case of K.C.: contributions of a memory-impaired person to memory theory. *Neuropsychologia* 43:989–1021.

Rosenbaum, R. S., D. T. Stuss, B. Levine, and E. Tulving. 2007. Theory of mind is independent of episodic memory. *Science* 318:1257.

Rothspan, S., and S. J. Read. 1996. Present versus future time perspective and HIV risk among heterosexual college students. *Health Psychology* 15:131–34.

Shah, A. K., S. Mullainathan, and E. Shafir. 2012. Some consequences of having too little. *Science* 338:682–85.

Snoek, A., N. Levy, and J. Kennett. 2016. Strong-willed but not successful: the importance of strategies in recovery from addiction. *Addictive Behaviors Reports* 4:102–7.

Vanderveldt, A., L. Oliveira, and L. Green. 2016. Delay discounting: pigeon, rat, human – does it matter? *Journal of Experimental Psychology: Animal Learning & Cognition* 42:141–62.

Zimbardo, P., and J. Boyd. 1999. Putting time in perspective: a valid, reliable individual-differences metric. *Journal of Personality and Social Psychology* 77: 1271–88.

Zimbardo, P. G., K. A. Keough, and J. N. Boyd. 1997. Present time perspective as a predictor of risky driving. *Personality and Individual Differences* 23:1007–23.

CHAPTER 6

In What Sense Are Addicts Irrational?

Howard Rachlin *

In this chapter I take a pragmatic look at rationality and addiction. What is the best way to use the concept of rationality when talking about addictive behavior? I view rationality in terms of overt behavioral patterns rather than as a smoothly operating logic mechanism in the head. I examine and reject the notion of rationality as consistency in choice – the property of exponential delay discount functions (which trace the decrease in value of a reward as its delay increases). Addicts are not irrational because of the type of delay discount function that governs their choices – or because of the relative steepness of that function. Addicts' delay discount functions are indeed steeper than those of nonaddicts (MacKillop et al. 2011; Odum 2011), and steepness of discounting indicates a tendency to prefer smaller, sooner over larger, later rewards. But, as I will argue later, such a tendency is not necessarily irrational. Instead, rationality may be seen as a form of *soft commitment* – creating patterns in one's own behavior so as to maximize value in the long run. Addicts are irrational to the extent that they fail to create such patterns.

6.1 Behavioral and Cognitive Viewpoints

In order to get a handle on what it means to behave rationally, let us look at the concept pragmatically. Instead of asking, "what does rationality really mean?" or "where is rationality located?" or "how does rationality work?" or "is this or that animal fundamentally a rational animal?" or "is this or that behavior really rational or really not rational?," it might be better to ask, "how should psychologists *use* the word *rational*?" or "what is its proper function in our scientific language?" Let us consider two fundamentally different ways to use the term.

* Preparation of this chapter was supported by Grant No. 1630036 from the National Science Foundation.

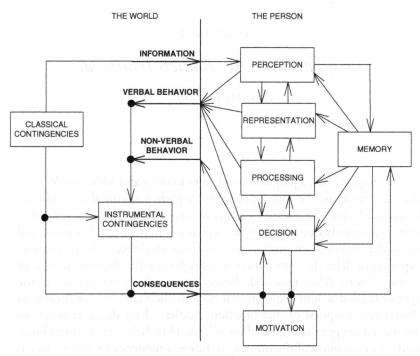

Figure 6.1 General outlines of a neurocognitive theory (right) and
a behavioral-economic theory (left)

Figure 6.1 illustrates a distinction between two ways of looking at
behavior and, by implication, two ways of using the expression *rational
behavior*. The thick vertical line divides the inside of a person (represented
on the right) from the external world (represented on the left). On the right
is my own version of a generic cognitive model of decision making [a rough
version of Kahneman and Tversky's (1979) prospect theory]. I mean it to
illustrate a *kind* of model and not any particular model. Looked at from the
right, information from the outside world comes into the person through
the sense organs and enters the cognitive mechanism; the information
proceeds through a series of submechanisms; it is perceived, represented
(or encoded), processed, and used in making a decision and eventually an
overt choice. These submechanisms are affected directly or indirectly by
memory, by feedback from the external world, and from below, by motivational variables such as hunger, thirst, and other fundamental or not-so-
fundamental drives. Each of the submechanisms may, in turn, be divided
into sub-submechanisms, and so forth. Cognitive theorists may infer the

states of the mechanisms from either nonverbal choice behavior or verbal reports of those states.

There is some dispute among philosophers of mind whether the lines of division in any cognitive theory correspond to, or may be reduced to, neural mechanisms – that is, whether cognitive psychology is reducible to neuroscience (Churchland 1989) – or whether cognitive psychology and neuroscience each carve out nonoverlapping units within the nervous system (Dennett 1989). The *language* of cognitive psychology does overlap with the *language* of neuroscience to some extent. Our mental vocabulary, including terms such as *memory, perception,* and so forth, is common to them both. I therefore call the model on the right side of the diagram a *neurocognitive model.* Cognitive scientists and neuroscientists both have the same goal – to discover mechanisms within the organism (Gazzaniga 1998).

On the left side of the diagram is what I call the *behavioral-economic model.* This model uses the very same inputs and outputs as the neurocognitive model, but the boxes stand not for *spatially defined mechanisms* but for *temporally defined contingencies.* A *classical contingency* is a relation between two environmental events – the bell and the food powder (in the case of Pavlov's dogs) or the train whistle and the train. An instrumental contingency is a relation between behavior and consequences. A fixed-ratio schedule, which says that a rat will obtain a food pellet after every 10 lever presses; the price of a loaf of bread, which says that if you give the baker so much money, he will give you so much food; and the conflicting relations between smoking and feeling good and between smoking and lung cancer are all instrumental contingencies.

To contrast the neurocognitive viewpoint with the behavioral-economic viewpoint, consider the question, "why am I sitting here at my computer and typing?" An email came to me several months ago; stimulated my eyes; was perceived, represented, and processed; and activated a decision mechanism. The output of that mechanism was a wholly internal decision to write this chapter. Then that decision activated certain motor centers and got me moving. I wrote the due date on my calendar, but I also encoded it, or a fuzzy version of it, in my memory. I responded to the various emails. Then, when my calendar indicated that the due date was rapidly approaching, my lower motivational processes – what Loewenstein (1996) calls "visceral" processes – became activated. I began to write. Every word I am writing may be seen in terms of the operation of the neurocognitive mechanism inside me. There is no question that some version of this model must be true. No organism, not even an amoeba – much less a person – is empty.

Yet there is another way to look at the relation among the heavy arrows of the diagram – from the outside rather than from the inside. For instance, there is another way to look at the question of why I am writing this chapter. You could say that I am writing because I *hope* to influence the behavior of readers or at least to convince them to look more kindly than they already do on behavioral research. Or I am writing it because the act of writing down my thoughts will help me to develop my ideas, or because I *believe* that writing this chapter will somehow further my career.

The boxes on the left side of the diagram look like the boxes on the right, but they stand for radically different entities. The boxes on the right stand for current states of currently existing mechanisms. You could, in theory, point to them in the same way that you could point to a car's carburetor. To the extent that the boxes on the right are hidden, they are hidden in space, somewhere inside the organism. In contrast, the boxes on the left stand for temporally extended contingencies, that is, relations over time between patterns of behavior and environmental events. You could not point to such contingencies any more than you could point to the relation between the force of your foot on a car's accelerator and the speed of the car. I call this viewpoint (from the left in Figure 6.1) *teleological behaviorism* (TEB).

From the viewpoint of TEB, terms such as *hope* and *belief* stand not for internal mechanisms but for patterns of behavior extending over perhaps long time periods. So far we have neither neurocognitive nor behavioral theories of hope and belief. How would a behavioral-economic model treat a term such as *perception*? Here is a discussion of perception from Rachlin (2017: 67–68):

> For TEB, perception is identical to a correlation over time between a person's overt behavior and an identifiable pattern of events in the environment. Consider the following question: What is the difference between two people (say John and Marcia), one of them (Marcia) stone deaf, both sitting stock still while a Mozart quartet is playing? A. John is hearing (i.e., perceiving) the music whereas Marcia is not hearing it. Q. What does it mean to hear? A. To discriminate by overt acts, over a period of time, between sounds and silence. That is, a non-zero correlation exists between John's behavior and sounds (unsignaled through other senses) whereas there is no correlation (a zero correlation) between Marcia's behavior and such sounds. During the past, in the presence of sound signals, their behavior (perhaps including taking audiometric tests) differed, and will differ in the future. (Consider their differing reactions to someone rushing into the room behind them yelling, "Fire!") Their identical behavior during the Mozart quartet is merely one congruent point in two drastically different correlations between behavior and sound.

It could be that Marcia's hearing mechanism is entirely normal but she is nevertheless unresponsive to sounds. In that case we would say she was "psychologically deaf." Is psychological deafness real deafness? From the viewpoint of TEB, the answer must be "Yes." What counts for deafness as for all psychological (or mental) states, for TEB, is Marcia's behavior in the long run. If she was faking deafness, then her subsequent behavior would reveal what her state really was. If, despite her normal hearing mechanism, she continued to behave all her life as a deaf person behaves, the question: "Was she faking deafness or psychologically deaf?" would be entirely non-pragmatic.

6.2 Rational Behavior and Addiction from a Neurocognitive Viewpoint

How, then, may the concept of rationality or rational behavior be used in cognitive theories? For a cognitive theory, rational behavior is usually seen as the product of a smoothly functioning, unimpeded, logical decision mechanism. Different neurocognitive theories would have different ideas of how that machine works (i.e., differing *normative* models) and of how it may be impeded. If, as is often the case, the predictions of a given neurocognitive theory are disconfirmed, the theory may be modified to explain the discrepancy. For instance, a nonlinear relation between probability and decision weight may be assumed, as in Kahneman and Tversky's (1979) prospect theory, or hyperbolic time discount functions may be used instead of exponential time discount functions to predict choice. These changes may be viewed as changes in perception or changes in processing rather than as changes in logic, which remains the same as it was in the original theory – the unimpeded operation of an internal logical mechanism. Now, after it has been explained, behavior that was labeled as irrational in the original theory may be labeled rational. This doesn't mean that people have suddenly become smarter than they were before, just that the label applied to their behavior has changed. You might say that the behavior is not so much *rational behavior* as *rationalized behavior*. Or the explained behavior may continue to be viewed in terms of a rational core affected by a catalog of heuristics or biases.

Nowhere in this cognitive-physiological conception of rationality (as interpreted by a noncognitivist, nonneuroscientist) is the requirement that rational behavior also be conscious behavior. A person might consciously deliberate before acting and still act irrationally, or act first and then rationalize her own behavior afterwards. or have no conscious awareness of her behavior at all and yet act rationally. Consciousness, whatever it might be (and it might or might not be a useful concept at all within any cognitive theory), seems orthogonal to neurocognitive rationality.

Given this very general cognitive view of rationality, how would it apply to addiction? One obvious way is to say that whereas the addict may possess a perfectly adequate logic mechanism, the operation of that mechanism may be impeded by immediate motivational forces acting from below (*visceral factors*). When such forces are strong enough that they may overwhelm the output of the logic mechanism and produce irrational behavior, addictive behavior may then be seen as one form of irrational behavior. If, at a moment when the motivational forces are inactive, you ask a nicotine addict, for example, whether she wants to be a nicotine addict, she says no; this is her logic mechanism working in an unimpeded way. However, when the motivational forces are strong enough – when she is offered a cigarette – she smokes it.

An addict may differ from a nonaddict in two ways: (1) the nonaddict's motive to consume the addictive substance may be weaker than that of the addict ("natural virtue" or "temperance," according to Aristotle) or (2) the output of the nonaddict's logic mechanism may be stronger or less resistant to disruption than that of the addict ("continence," according to Aristotle – what we would call *self-control*). So, when my wife, who hates chocolate, refuses the chocolate dessert, she's being naturally virtuous but not self-controlled; when I refuse it, in contrast, I'm controlling myself. (Of course, a person may have both natural virtue and self-control.) The concept of *rational addiction* would be self-contradictory from the cognitive-physiologic viewpoint because addiction, from that viewpoint, is irrational. This, then, is one of many possible views of addiction from a neurocognitive viewpoint – the right side of Figure 6.1.

6.3 Rational Behavior and Addiction from a Behavioral-Economic Viewpoint

Here is a quotation from the comedian Dick Cavett (*New York Times*, Week in Review, May 31, 2009, p. 10.): "Once, as [Cavett and Jonathan Miller of 'Beyond the Fringe'] waited backstage together at the 92nd Street Y in New York City, [Cavett] pointed disapprovingly at [Miller's] lit cigarette. [Miller said,] 'I know these will kill me, I'm just not convinced that this particular one will kill me.'" Miller is right. No particular cigarette can harm a person, either now or later. Only what is essentially an abstraction – the relation between rate of smoking and health – will harm him.

We all tend to focus on the particular when it comes to our own behavior. Only when, like Cavett, we observe someone else's behavior, or when circumstances compel us to experience the long-term consequences of our own behavior, are we able to feel their force. Another way of putting Miller's problem is to say that Miller's behavior was under the control of the consequences of smoking the particular cigarette, whereas it should have been under the control of the consequences of smoking at a high rate over a long period of time. Can we say that Miller is just not a very rational person, perhaps not capable of understanding abstract concepts? No way. Here is his Wikipedia entry:

> Sir Jonathan Wolfe Miller CBE (born 21 July 1934) is a British theatre and opera director, actor, author, television presenter, humourist, sculptor and medical doctor. Trained as a doctor in the late 1950s, he first came to prominence in the early 1960s with his role in the comedy revue Beyond the Fringe with fellow writers and performers Peter Cook, Dudley Moore and Alan Bennett. He began directing operas in the 1970s and has since become one of the world's leading opera directors with several classic productions to his credit. His best-known production is probably his 1982 "Mafia"-styled Rigoletto set in 1950s Little Italy, Manhattan. In its early days he was an associate director at the Royal National Theatre and later he ran the Old Vic Theatre. He has also become a well-known television personality and familiar public intellectual in both Britain and the United States.

Molar behavioral theories and economic theories of individual behavior (microeconomic theories) take the same form. The instrumental contingencies of behavioral theories correspond to the constraints (prices and budgets) of economic theories. Both specify relations between behavior and consequences. The behavioral concepts of reinforcement maximization and matching correspond to the utility functions of economic theory (Rachlin 1989).

How does the concept of rationality fit into microeconomics? Economic utility theory assumes that people behave so as to maximize utility under any given set of constraints (prices and budgets in economic language – contingencies or reinforcement schedules in behavioristic language). The object of economic theory is to discover the utility function that is maximized. Once such a function is discovered under one set of constraints, it may then be tested under another set of constraints. If the same utility function is maximized under both sets of constraints, it may then be tested under a third set of constraints, and so forth – until it fails a test, as it inevitably must. At that point, the utility function is modified or parameters are added to it until it describes behavior under all tested constraints, and the process continues – at least in theory. The result, in theory, is a grand utility function that is maximized under all possible sets of constraints. This is the method of *revealed preference* (Samuelson 1973). Obviously, the desired end will never be reached – just as no perfect neurocognitive model will ever be developed. But, in the process of development, economic theory is supposed to become better and better able to predict behavior under any imposed set of constraints. The utility functions are methods the theorist uses to predict a person's behavior in one situation from his behavior in other situations. Of course, there are internal mechanisms underlying all behavior, but there may be no internal mechanism corresponding to any particular utility function. To use an example from Dennett (1978), a chess player may reliably bring out her queen too soon – that is, bringing out her queen too soon may be an accurate description of her past behavior and useful in predicting her future behavior – but this tendency may not be encoded as such in any specific mechanism inside of her head.

It is often the case that based on a given utility function, behavior that maximizes utility in the *relative* short run does not maximize utility, as measured by that particular function, in the *relative* long run. If there are two people, one of whom maximizes utility in the long run (hence not in the short run) and one of whom maximizes utility in the short run (hence not in the long run), the behavioral economist could say that the first person's behavior is (relatively) rational and the second person's behavior is (relatively) irrational. A tobacco addict would be a good example of a person of this second kind. By smoking, the addict maximizes utility in the short run (the person facing a firing squad might as well have a smoke) but not in the long run – in terms of health, social acceptance, expense, and so on.

However, suppose that an economic theory claims to have discovered a single utility function that describes both the addict's and the nonaddict's behaviors (e.g., Becker and Murphy 1990). That is, given a set of parameters within the model, the addict may be seen to be maximizing utility in the long run. The economic theory would then have rationalized the addict's behavior. As in the neurocognitive theory, this does not mean that people have suddenly become smarter than they were before. It just means that the theory has developed so that by means of varying the parameters of a single utility function, the theory can explain and predict both the addict's and the nonaddict's behaviors. A certain consistency is discovered in the addict's behavior that seemed not to have been there before.

In Rachlin (1989), I tried to do just that with respect to Tversky and Kahneman's (1981) cognitive decision theory and economic maximization.

From the behavioral-economic viewpoint, as from the neurocognitive viewpoint, rationality has nothing to do with consciousness. Rational addiction means that the addict's behavior, as well as the nonaddict's behavior, maximizes utility in the long run according to some particular utility function. It does not mean that the future addict sits down on the day of his bar mitzvah and plans out the rest of his life.

6.4 Crossing Discount Functions and Rationality

Consider the practice of compounding interest by banks. With compounding, interest is calculated (at a fixed rate) over some fixed period t, added to the principal, and repeated at intervals of t. If, when you came to withdraw your money, the bank had instead calculated simple interest from the time of deposit, you would have an incentive, after a short period, to withdraw your money plus the interest and deposit it in another bank, thus compounding the interest yourself. So as not to lose your account in this way, the bank compounds your money for you. As the period of compounding t approaches zero, your money would be compounded at an infinite rate, and the resulting overall discount function would approach the exponential function

$$\frac{v}{V} = e^{-it} \tag{6.1}$$

where V is the balance for an original deposit of v after a time t and with an interest rate i.

When depositing money and earning interest, the more the delay until you withdraw the money, the greater is the accumulated amount, and the more that amount is worth to you. Thus the value of that reward *grows* with time. The lowercase v represents the smaller amount deposited; the uppercase V represents the larger amount accumulated.

Now suppose that a delayed reward was discounted using this same function (exponential delay discounting). Then the more the delay until you get the money, the less it is worth to you now. The value of the delayed reward *diminishes* with delay; the uppercase V now represents the larger, undelayed, amount; the lowercase v represents the discounted value. The fraction v/V is the (normalized) degree of discounting, t is the delay, and interest rate i is a constant representing the degree of discounting. With higher values of i, discounting would be steeper. That is, the higher the value of i, the less a given delayed reward is worth. If the choices of two people in a given situation were described by a delay discount equation, such as (6.1), and the function describing the choices of one of them had a higher i value than the other, that person would be expected to be more impulsive in her choices – would tend to choose smaller, sooner rewards over larger, later rewards to a greater extent – than the person with a lower i value. A theory of addiction based on exponential delay discount functions would say that addicts have higher i values than nonaddicts. However, addicts would be no less rational (according to the economist's definition) than nonaddicts – as long as their choices were consistent.

In what sense does exponential time discounting (6.1) imply consistent choice? With exponential delay discounting, two delay discount functions with the same interest rate (i in (6.1)) would not cross (Ainslie 1992). If, contrary to fact, exponential delay discounting described actual human and nonhuman choices among delayed rewards, and if a person preferred \$100 delayed by 10 days to \$95 delayed by 9 days, then, after 9 days had passed, that person would still prefer the \$100, now delayed by a day, to the \$95 available immediately; the discount functions for the \$100 and the \$95 would not cross. Nevertheless, in many instances, people do change their preferences over time, preferring the larger, later reward when both rewards are relatively distant but switching their preference to the smaller, sooner reward when it becomes imminent. That is, people's actual discount functions may indeed cross. Does this mean that their choices are irrational?

Although it is common for *banks* to approximate exponential discounting by compounding interest over the period of a loan when borrowing or lending money for an indefinite period, it is not common for an *individual* to compound the appreciation of value continuously over fixed delays when choosing among delayed rewards. Indeed, there is strong evidence that people's choices are not well described by the exponential discount function of (6.1). The choices of people (as well as nonhuman animals) have been found to conform instead to *hyperbolic discounting*.

With hyperbolic delay discounting,

$$\frac{v}{V} = \frac{1}{1 + kD} \tag{6.2}$$

where v/V is the degree of discounting measured as a fraction, D is delay to the reward (corresponding to t in (6.1)), and k is a constant, measuring the degree of discounting (corresponding to i in (6.1)). In virtually all experimental determinations of psychological discounting, (6.2) (or a variant with the denominator exponentiated) is the form empirically obtained (Green and Myerson 2004).

Reconsider the preceding example in which a person is assumed to prefer $100 delayed by 10 days to $95 delayed by 9 days. With (6.2) (and a sufficiently high value of k), after 9 days had passed, that person would now prefer the $95 available immediately to the $100 now delayed by a day. In other words, unlike exponential delay discount functions with the same interest rate i, hyperbolic discount functions with the same discount rate k may cross. Despite the reversal in preference as time passes, there is nothing necessarily irrational about hyperbolic discounting per se, even by forward-looking organisms. That is, crossing functions are not in themselves necessarily irrational.

The hyperbolic discount functions of two- and three-dimensional energy propagation may cross under conditions corresponding to crossing delay discount functions. For example, the sound energy from an iPhone close to your ear may be more intense than that of a moving subway train that is 10 feet away, but stepping back 10 feet – now you'd be 10 feet from the iPhone and 20 feet from the train – the sound energy intensities of the two would reverse. But there is nothing irrational about this purely physical process.

What may be considered irrational, however, is a failure to account for a change of mind when it is known that a second choice will be offered and, based on past experience, that one's own preference will reverse (O'Donoghue and Rabin 1999). The behavior of an alcoholic who vows, during a morning hangover, never to drink again and then goes to a party, when in the past he has always gotten drunk at such parties, may be labeled as irrational – not because the addict changed her mind about drinking when she got to the party but because the addict failed to anticipate that she would change her mind. To put it another (and more abstract) way, the addict's behavior is controlled by the contingencies of the moment; the nonaddict's behavior is controlled by contingencies operating over a wider time span. The remainder of this chapter is essentially a development of this point.

6.5 Rationality In Social Behavior

The following passage from Anthony Trollope's novel, *The Way We Live Now* (1875/1982, Oxford University Press) presents an analogy between a character's selfishness and his impulsiveness (p. 17):

> Whether Sir Felix ... had become what he was solely by bad training, or whether he had been born bad, who shall say? It is hardly possible that he

should not have been better had he been taken away as an infant and subjected to moral training by moral teachers. And yet again it is hardly possible that any training or want of training should have produced a heart so utterly incapable of feeling for others as was his. He could not even feel his own misfortunes unless they touched the outward comforts of the moment. It seemed that he lacked sufficient imagination to realize future misery though the futurity to be considered was divided from the present but by a single month, a single week, – but by a single night.

Trollope here attributes Sir Felix's selfishness, his social narrowness, to his lack of self-control, the narrowness of his time horizon. Is there a relation between rationality in self-control and rationality in social cooperation?

Jones and Rachlin (2006) and Rachlin and Jones (2008) found that (6.2) precisely described the average results of human participants who each chose whether to share a hypothetical amount of money with another person at a greater or lesser social distance. The discounting variable, *social distance N*, was measured as numerical order in closeness, to the participant, of the person who would be sharing the money (#1 being the closest person, #2 being the second closest, and so forth). Participants were asked to imagine that they had made a list of the 100 people closest to them in the world ranging from their dearest friend or relative at #1 to (possibly) a mere acquaintance at #100.

Then the participants were asked to think of a person on the list (#20, for example) and to choose (hypothetically) between $75 to be given to that person and (initially) $75 for themselves (for half of the participants) or $0 for themselves (for the other half). Almost all participants chose $75 for themselves over $75 for #20 and $75 for #20 over $0 for themselves. Then we lowered the participant amount stepwise from $75 to $0 for half the participants and raised the participant amount stepwise from $0 to $75 for the other half while keeping the amount for the receiver of the money (#20 in this case) constant at $75. Almost all participants reversed their preference at some point in the sequence. This point of reversal is the maximum amount of money the participant would be willing to forgo to give $75 to the receiver (#20 in this case). For instance, a participant may have preferred $40 for herself to $75 for person #20 and preferred $75 for person #20 to $30 for herself. In this case, the crossover point would be taken as $35 – the maximum amount the participant was willing to forgo to give $75 to #20 on her list. Then we repeated the procedure for other social distances in random order. (Those tested were $N = 1, 2, 5, 10, 20, 50,$ and 100.)

We found that the greater the social distance of the receiver from the participant N, the less money the participant was willing to forego. That is, "generosity" was discounted by social distance according to the hyperbolic formula

$$\frac{v}{V} = \frac{1}{1 + kN} \qquad (6.3)$$

where v is the point of reversal. With social distance N taking the place of delay D in (6.2) and $k = 0.05$, the variance R^2 accounted for by (6.3) was 0.997. Thus delay and social discounting are found to take the same mathematical form.[1]

Equation (6.3) was fit to the median crossover points of each of the participants. Fitting the equation to individual participants, the constant k varied over a wide range. Just as a high delay discounting constant k in (6.2) implies lack of self-control, so a high social-discounting constant k in (6.3) implies lack of generosity. There is a significant positive correlation (among college students) between reported number of cigarettes smoked per day by an individual and that individual's social-discounting constant (Rachlin and Jones 2008). Moreover, individual delay discounting was significantly correlated with social discounting over individual participants. That is, people whose discount functions implied a tendency to be self-controlled also tended to be generous to others. Why might this be so?

As Ainslie (1992) and Rachlin (2000) point out, you can view an individual over time analogously to a series of individuals over space.[2] Again, let us look to the arts for an example – the *Seinfeld* TV show this time. In one of his introductory routines (as I remember it), Seinfeld talks about "night Jerry" and "day Jerry." Night Jerry has fun; he stays out late, gets drunk, spends money. Day Jerry is the one who must wake up early in the morning with a painful hangover. Night Jerry has only contempt mixed with pity for day Jerry, whereas, of course, day Jerry hates and resents night Jerry. But what can day Jerry do to get even? His only recourse, Seinfeld says, is to stay in bed late, perform badly at his job, and then get fired; then night Jerry won't have enough money to go out and get drunk. Jerry's problem is very much like Sir

---

[1] Social distance N is an ordinal measure and by itself not a meaningful scale. However, Rachlin and Jones (2008) found that N varies as a power function (linear in logs) of subjective physical distance, a ratio scale. We nevertheless continue to use N as an independent variable in these experiments because it is easier for participants to make judgments based on N than based on physical distance.

[2] See other contributions to this volume, especially those of Arif Ahmed (Chapter 4) and Natalie Gold (Chapter 10), for different approaches to this same conclusion. The problem of self-control is essentially a conflict between self-interest over a narrow span of time and self-interest over a wide span of time.

Felix's problem in Trollope's book. Both characters fail to bring their behavior under control of long-term contingencies – their future selves. Their delay discount functions are very steep because they see their own selves at a future time as socially distinct from their own selves at the present moment. Addicts, I would argue, are in such a position.

Addicts treat their future selves as they would treat people far from them in social distance – people unlikely to reciprocate any cooperative behavior they may exhibit. If, in future interactions, other people are unlikely to cooperate with you, regardless of your behavior now, it would be foolish to cooperate with them now. To take an example from Aristotle, an addict is in the position of a soldier in a rout in battle who, if he turns and makes a stand, will be ignored by the other fleeing soldiers. The irrationality of addicts may be most usefully understood to lie not in the description of their choices by one sort of discount function or another but in their failure to identify with their past and future selves – their failure, that is, to distinguish between their own past and future selves, separated from them in time, and other people, separated from them in social space. That is, even when she is nominally playing a game against her future self, the addict plays it as a nonaddict would play a game against other people, people socially distant from her.

6.6   Soft Commitment

A pigeon is faced with a choice between pecking two buttons. If it pecks one button (say the left one), it receives one food pellet immediately (the smaller, sooner reward). If it pecks the other button (the right one), it receives four food pellets delayed by four seconds (the larger, later reward). In such a situation (after repeated exposure to the choice), the pigeon will come to peck the left button (smaller, sooner reward) nearly 100 percent of the time. The same pigeon, however, will peck the button leading to the larger, later reward if each reward requires 20 pecks rather than a single peck – even though the pigeon is free to switch after the nineteenth peck (Siegel and Rachlin 1995). This is significant because after 19 pecks, the pigeon is in the same position as when it was required to make only one peck. Why does the pigeon, for its final peck, persist in its prior choice and not switch to the smaller, sooner reward button, the button that it would clearly prefer if it did not have to make those 19 prior pecks? Yet pigeons do begin the 20 pecks on the larger, later reward button, and do persist in their prior choices, and do obtain the larger, later reward.

We call such response persistence *soft commitment*. In this case, the pigeon's *fixed-ratio* pattern of pausing and then rapidly running off the 20-peck requirement quadrupled its overall reinforcement rate. For pigeons, this patterning is an innate response to the 20-peck requirement. But humans may learn to institute such patterns in their own choices between smaller, sooner and larger, later rewards.

The following experiment (Kudadjie-Gyamfi and Rachlin 1996) tests the effect of patterning on self-control (maximization) with human participants. The version of the experimental game used gave a single point (convertible to cash at the rate of 10 cents per point) for Y choices and X choices alike but varied *delays* of point gain by choice of different alternatives. The total cumulative session time was fixed (at 325 seconds). Participants chose by pressing buttons marked A (functioning as Y) and B (functioning as X). A computer screen displayed the remaining session time and total points earned. After pressing either button, the participant waited for a certain delay period while the session timer on the computer screen counted down. At the end of the delay, the timer stopped counting, and one point was added to the participant's displayed score. When the session timer reached zero, the experiment ended. The participants would maximize total reward by minimizing average delay. The rules (not revealed to the participants) were the following:

1. Each choice of Y yields 1 point delayed by $(N + 3)$ seconds.
2. Each choice of X yields 1 point delayed by N seconds.
3. N at each choice is equal to the number of Xs in the previous 10 choices.

Figure 6.2 shows the contingencies graphically. Suppose that over the previous 10 choices, 5 were for X and 5 for Y. Then, as Figure 6.2 indicates, if the next choice were X, the delay would be 5 seconds; if the next choice were Y, the delay would be 8 seconds. The parallel lines show that regardless of prior choices, X-choice delays are always 3 seconds less than Y-choice delays. However, if the participant keeps making X choices, both delays would increase until at last all delays would be 10 seconds, whereas if the participant made only Y choices, all delays would be 3 seconds, more trials would fit into the session, and more points would be earned.

Again, the *particular* consequences of choosing X were better than those of choosing Y (3 seconds less delay), whereas the *general* consequence of choosing X was to increase N, thereby increasing average delay. Of the 60 participants in this experiment, none could verbalize the rules after the experiment, although almost all understood that it was sensible to at least

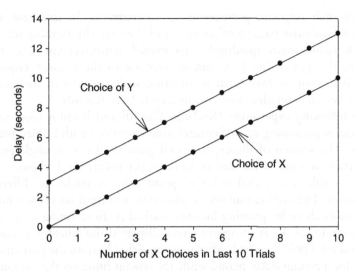

Figure 6.2 Contingencies of the Kudadjie-Gyamfi and Rachlin (1996) experiment

occasionally choose *Y*. Most distributed their choices between *Y* and *X*. There were four groups of participants differing only with respect to the patterning of trials. All participants played under the set of rules stated earlier. The trial patterns of the four groups were as follows:

> Control group 1: ... *COCOCO* ...
> Control group 2: ... 10*sCO* 10*sCO* 10*sCO* ...
> Control group 3: ... 30*sCO* 30*sCO* 30*sCO* ...
> Experimental group: ... 30*sCOCOCO* 30*sCOCOCO* 30*sCOCOCO* ...

where *C* = choice, *O* = outcome, and 10*s* and 30*s* represent intertrial intervals (the session timer did not count down during those intervals). The experimental group was the only one with patterned trials – triples of rapid trials separated by 30-second intervals. The patterning, it was hypothesized, would group trials into threes and emphasize the consequences of groups of trials instead of specific individual trials. Relative to the experimental group, control group 1 had the same local rate of trials, control group 2 had the same overall rate of trials, and control group 3 had the same intertrial interval. Figure 6.3 shows the results of this experiment.

The experimental group chose *Y* significantly more times than did any of the control groups, while control group 1 (with a 0 intertrial interval) chose *Y* significantly fewer times. (Money earned was directly proportional to percent of *Y* choices.) Analysis of choices within the grouped triples of

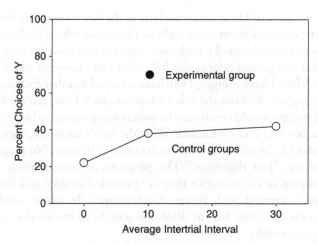

Figure 6.3 Results of the Kudadjie-Gyamfi and Rachlin (1996) experiment

the experimental group indicates that the probability of a *Y* choice in the first of the three grouped trials was 0.48. However, if a *Y* choice was made on the first trial, the probability of a *Y* choice on the second trial of the three contiguous trials was 0.54. And if the first two were *Y* choices, the probability of a *Y* choice on the third trial was 0.80. Thus we found a tendency to persist in a choice leading to a larger long-term reward once that choice had been made initially.

Why does soft commitment increase self-control? When we commit ourselves to a behavioral pattern, we are reducing our future options and hence the potential variability of our future behavior. As soon as we embark on a particular behavioral pattern, we have abandoned all other potential patterns. The difference between the prisoner and the free person is that the free person may potentially do what the prisoner can do – *plus* other things. A pigeon presented with the choice between a smaller, sooner reward and a larger, later reward may consistently and monotonously prefer the former. If at an earlier time, however, the pigeon had committed itself to the larger, later reward, it would have reduced the potential, if not the actual, variability of its behavior (Rachlin and Green, 1972). In other words, it would have reduced its own freedom. Commitment *means* reduction of freedom, and freedom *means* potential behavioral variability. Thus commitment *means* reduction of potential behavioral variability.

In other words, soft commitment adds "weight" to behavior; it focuses attention on the behavior's long-term consequences. Consider the person who keeps a supply of Dove ice-cream bars in the freezer and is in the habit of eating one or two of them every night or the person who is in the habit of having two or three glasses of scotch every night before going to sleep. Now suppose that the person adopts the following rule: however many Dove bars (or scotches) I have tonight, I will have an equal number each night for the next 25 nights. Without the rule's adoption, each Dove bar (or scotch) consumed tonight would entail only its own consequences and could be, in theory, the last one ever consumed. With the rule's adoption, each Dove bar (or scotch) is, in effect, 25 of them strung out in time. No longer is it possible to say, "Just this once." The pleasures of consumption are no longer restricted to the moment; they are extended in time and therefore more easily compared with future disadvantages. In other words, the decision to eat the Dove bar, to drink the scotch, to smoke the cigarette has been given weight.

6.7 Implications for Treatment of Addiction

Although the purpose of this chapter is to outline and discuss two viewpoints (neurocognitive and behavioral economic) of the concept of rationality in addiction and not to suggest treatment methods, it may be clarifying to consider how the two viewpoints result in two different approaches to treatment. The neurocognitive viewpoint would focus on the two internal submechanisms of addiction – the logic mechanism and the motivational interference with that mechanism. Current neurobiology has given us hints as to where in the brain these internal mechanisms and interactions are located, but we know very little about how they work. Moreover, in its focus on internal mechanisms, the neurocognitive approach tends to ignore behavioral context.

The behavioral-economic viewpoint suggests training addicts in self-control by increasing the temporal extent of patterns of alternative behavior. Learning an effective personal rule that may come to control and reduce addictive behavior is not primarily a cognitive problem. It is easy enough for a person to learn to repeat a rule. The difficult task is to bring behavior under control of that rule as a signal for valuable nonaddictive patterns. (See the short horizontal arrow pointing to the box labeled "INSTRUMENTAL CONTINGENCIES" in Figure 6.1.) For this task, behavioral methods may be superior to neurocognitive ones. Nevertheless, pursuit of both neurocognitive and behavioral-economic treatment approaches would seem to be the most fruitful tactic for treatment.

6.8 Rationality and Self-Control in Everyday Life

Consider the person who, attempting to be rational in the neurocognitive sense, decides each time he pays a restaurant bill whether or not to leave a tip. Not only will that person have to pay attention to the quality of the service and the probability of his coming back to that particular restaurant, and to process that information, but he will have to consider the probability that faced with adding a not-inconsiderable sum of money to his bill, "visceral factors" might influence his reasoning and bias him in the direction of leaving less (or perhaps even more) than the optimal amount. I am not aware of research on the subject, but my guess is that when evaluating the amount of money to be spent at some future time in some specific future situation, most people will calculate an optimal amount different from the amount they calculate should be spent in that very same situation right now. In other words, *visceral factors* (acting currently but not on contemplation of the future) influence not only our motives but also our reasoning.

Realizing that our on-the-spot decisions are often biased in this way, we allow our behavior to conform to some molar pattern that we have found to be generally optimal – always leave a tip (or never leave a tip, or leave 15 ± 2 percent depending on how we feel, if that's what we have determined). Indeed, such behavior, accounting as it does for our own frailty, may usefully be labeled as more rational than that based on a more precise calculation of current contingencies that fails to account for that frailty. Similar considerations may govern other social activities such as contributing to public broadcasting or to charities, voting, not littering, busing your tray in a cafeteria, and so on.

Just as it may be rational in social-cooperation situations to obey general rules – to make decisions on a global rather than on a local basis – it is rational to obey general rules in self-control situations. The addict may label her own behavior as rational when she lights up one "last" cigarette or drinks one "last" drink; after all, there will be great pleasure and virtually no harm in it if this one is indeed the last. But, as we all know, this is just the opposite of rational behavior – especially for an addict. To achieve self-control, addicts and nonaddicts alike must avoid making each decision on an apparently rational case-by-case basis and learn to make decisions in conformity with optimal molar (long-term) patterns (Rachlin 2000).

References

Ainslie, G. 1992. *Picoeconomics: The Strategic Interaction of Successive Motivational States within the Person*. New York, NY: Cambridge University Press.

Becker, G. S., and K. M. Murphy. 1988. A theory of rational addiction. *Journal of Political Economy* 96(4):675–700.

Brown, J., and H. Rachlin. 1999. Self-control and social cooperation. *Behavioural Processes* 47:65–72.

Churchland, P. S. 1989. *Neurophilosophy: Toward a Unified Science of the Mind-Brain*. Cambridge, MA: MIT Press.

Dennett, D. 1978. *Brainstorms: Philosophical Essays on Mind and Psychology*. Montgomery, VT: Bradford Books.

Dennett, D. C. 1989. *The Intentional Stance*. Cambridge, MA: MIT Press.

Gazzaniga, M. S. 1998. *The Mind's Past*. Berkeley, CA: University of California Press.

Green, L., and J. Myerson. 2004. A discounting framework for choice with delayed and probabilistic rewards. *Psychological Bulletin* 130:769–72.

Jones, B., and H. Rachlin. 2006. Social discounting. *Psychonomic Bulletin and Review* 17:283–86.

Kahneman, D., and A. Tversky. 1979. Prospect theory: an analysis of decisions under risk. *Econometrica* 47:263–91.

Kudadjie-Gyamfi, E., and H. Rachlin. 1996. Temporal patterning in choice among delayed outcomes. *Organizational Behavior and Human Decision Processes* 65(1):61–67.

Loewenstein, G. 1996. Out of control: visceral influences on behavior. *Organizational Behavior and Human Decision Processes* 65:272–92.

MacKillop, J., M. Amlung, L. Few, et al. 2011. Delayed reward discounting and addictive behavior: A meta-analysis. *Psychopharmacology* 216:305–21.

O'Donoghue, T., and M. Rabin. 1999. Doing it now or later. *American Economic Review* 89:103–24.

Odum, A. L. 2011. Delay discounting: I'm a k, you're a k. *Journal of the Experimental Analysis of Behavior* 96:427–39.

Rachlin, H. 1989. *Judgment, Decision, and Choice: A Cognitive/Behavioral Synthesis*. New York, NY: Freeman.

Rachlin, H. 1992. Teleological behaviorism. *American Psychologist* 47:1371–82.

Rachlin, H. 2000. *The Science of Self-Control*. Cambridge, MA: Harvard University Press.

Rachlin, H. 2017. In defense of teleological behaviorism. *Journal of Theoretical and Philosophical Psychology* 37:65–76.

Rachlin, H., and L. Green. 1972. Commitment, choice and self-control. *Journal of the Experimental Analysis of Behavior* 17(1):15–22.

Rachlin, H., and B. A. Jones. 2008. Social discounting and delay discounting. *Journal of Behavioral Decision Making* 21:29–43.

Samuelson, P. 1973. *Economics: An Introductory Analysis* (9th edn). New York, NY: McGraw-Hill.

Siegel, E., and H. Rachlin. 1995. Soft commitment: self-control achieved by response persistence. *Journal of the Experimental Analysis of Behavior* 64:117–28.

Tversky, A., and D. Kahneman. 1981. The framing of decisions and the psychology of choice. *Science* 211:453–58. Available at www.jstor.org.proxy.library.stonybrook.edu/stable/1685855.

CHAPTER 7

Why Temptation?

Chrisoula Andreou*

> It is often said that death and taxation are the only certain things in life. I think we can add temptation to this list. – A. T. Nuyen[1]

I. Consider a familiar case of temptation: bombarded by opportunities to eat sweets, an agent, call her *A*, eats them more often than she thinks she should (relative to her goals and how she ranks them). *A* assesses eating healthy food as very important, but she finds herself moved to eat sweets at a rate that conflicts with her assessment. Although she favors enjoying the *occasional* treat, she *routinely* opts for decadent delights and even recognizably mediocre treats, realizing a pattern of consumption that is clearly detrimental to her health. Given that *A* is proceeding in a way that is clearly inadequate for success by her own lights, the mismatch between her assessment (of the importance) of her goal and its motivating force figures as a failure of instrumental rationality and as a failure of rational self-control.

It might be supposed that an agent with a motivational system that is well suited for goal-directed agency would not find herself in such a predicament. Instead, her goals and the relative importance she assigns to each of them would either be determined by her motivations or determine them, precluding the possibility of a mismatch. My aim in this chapter is to suggest that susceptibility to temptation in familiar

* My thanks to Eric Hutton, Elijah Millgram, Anne Peterson, Sarah Stroud, Mariam Thalos, Chris Weigel, Peter Vallentyne, Mark D. White, Mike White, an anonymous reviewer, and audiences at the following venues for helpful comments on earlier drafts of this chapter: the Self-Control, Decision Theory, and Rationality Workshop; my work-in-progress talks at the University of Montreal and the University of Utah; the Time Bias and Future Planning Workshop; and the Centre for Time Conference on Time, Identity, and Future. I am also grateful for a supporting research grant from the University of Utah and for supporting research funds from the Tanner Humanities Center and the Sterling M. McMurrin Esteemed Faculty Award.

[1] Nuyen 1997: 102. Nuyen appeals to the idea that although "the perfectly good will is free of temptation ... such will is only ideal" (p. 97). As will become apparent, I explore an altogether different route to the conclusion that we are bound to experience temptation in our lives.

cases such as A's is not an anomaly but is instead integral to a type of motivational system that is generally quite well suited to the task of balancing multiple ongoing goals. The type of system that I will be examining has not been focused on in philosophical discussions of temptation or in related philosophical discussions of instrumental rationality. But, as I will explain, the features that characterize it make sense of many plausible but potentially puzzling views about temptation, including the following: (1) there is room for cases of giving into temptation that are not cases in which the agent's ranking of a tempting option changes as the tempting opportunity approaches and looms large, (2) although temptation is often heightened by visceral factors, it is possible to be tempted to overengage in activities that give no pleasure, and (3) temptation can threaten even if ultimate goals are not criticizable.

II.   Let me begin by emphasizing that I am not here concerned with the question of whether temptation is somehow integral to *free* agency or to *moral* agency. My concern is with the connection between temptation and *goal-directed* agency, whether or not such agency qualifies as the sort of full-blown agency necessary for freedom and morality.[2] On a related note, according to the notion of temptation at play in this chapter, temptation need not figure as a test of moral fortitude, and giving into temptation need not figure as a moral failing. Rather, the sort of temptation that is of interest here figures as a test of instrumental rationality, with failure understood as failure in relation to the agent's goals or the importance assigned to each of them.[3]

There are two dramatic variations on the case I opened with that should be distinguished from the pedestrian, but still potentially puzzling, sort of case that I seek to illuminate. In one variation, we have a "perverse case" in which the agent fails to "identify" with her assessment, "fully 'embrac[ing]' ... without compunction" a conflicting end,[4] such as, perhaps, the end of indulging her sweet tooth at every opportunity, come what may. In this variation on the case, the agent's failure to act in accordance with her assessment does not qualify as a failure of instrumental

---

[2] For some interesting contemporary contributions in this space, see, for example, Ainslie (2001), Bratman (1999, 2014), and Holton (2009).

[3] Importantly, I do not want to rule out the possibility that an ultimate goal may be criticizable (in general or for a particular agent at a particular time), in which case fallible but well-constituted agents may need to have ways of recognizing misguided goals and revising them. Such agents will need to have features that go beyond (and perhaps interact with) the features I will describe later. I leave this strand of inquiry for another occasion.

[4] The quoted phrases are from Watson (1987: 150).

rationality because the agent's assessment of the importance of eating healthy food does not play a role in determining what would count as success by the agent's own lights. In another dramatic variation on the case, the agent does not "fail to value what [she] judges valuable,"[5] but her behavior is, at times, hijacked by an "alien" or disowned desire, such as, perhaps, the desire to indulge her sweet tooth at every opportunity, come what may.[6]

In the mundane version of the case that is the object of my inquiry, the agent is not perversely alienated from her assessment and her control of her behavior is never highjacked by an alien desire. Still, because she is proceeding in a way that is clearly inadequate for success by her own lights, her control of her behavior does not qualify as rational. Although this version of the case does not appeal to a gap between what the agent values and what she judges valuable or to any alien desires, the mismatch that is appealed to is peculiar in its own way and calls for further illumination.

III. With these clarificatory points in mind, let us delve further into A's case: Suppose that, given a lounge with a large and steady supply of prominently displayed, mediocre donuts, A routinely finds that the "trivial good" of one such donut produces a disproportionately strong desire due, at least in part, to the "advantage of its situation" (including its visibility and close proximity).[7] Because A's motivational system does not preclude a mismatch between A's assessment of the importance of a goal and the goal's motivating force, A is easily led astray. It is, however, hasty to conclude that A is not well suited for goal-directed agency.

Granted, if we abstract from the challenges of effective choice over time and imagine time-slice agents (who persist for only a moment) with time-slice options (which can be completely carried out in a moment), a motivational system that allows for a mismatch between the importance an agent assigns to a goal and her motivation to pursue it does seem like an unadulterated curse. Suppose that A will be capable of action for only a brief moment t, during which she can either Q or R (both of which are valued by A). Insofar as we can think of A as ranking her options, the

---

[5] Ibid.

[6] The possibility of cases involving alien or disowned desires has already received substantial attention in the literature. For a seminal discussion concerning the ownership of desires, see Frankfurt (1971) and Watson (1975). For some relatively recent discussion in the same space, see, for example, Bratman (2007).

[7] The quoted phrases are from Hume (1978 [1739–40]), which I will return to later. Hume's nihilism about practical reason committed him to the view that even the persistent selection of a "trivial good" over "the greatest and most valuable enjoyment" is not "contrary to reason" (p. 416).

simplest, most effective motivational system – even if alternative systems are conceivable – will presumably prompt A to Q if Q-ing is more important to her than R-ing, to R if R-ing is more important to her than Q-ing, and to be indifferent between Q-ing and R-ing if neither is more important to her than the other.[8] But interesting complications arise when A is capable of action for more than a moment and A's multiple ongoing goals, though they pull in different directions, can each be accommodated to some extent over time. Consider, for example, the case in which A's goals include the pursuit of several different activities that cannot be engaged in simultaneously. As I will illustrate in the next two sections, in such cases, a motivational system that allows the motivating force of certain goals to be spurred or dampened even apart from any changes in the agent's assessment of the relative importance of each of her goals can serve the agent well by helping her efficiently balance her multiple ongoing goals.[9]

Importantly, I will focus on stretches of time during which A's (ultimate) goals and her assessment of their relative importance are fairly stable. During stretches of time when A's (ultimate) goals and her assessment of their relative importance are *not* fairly stable, A's pursuing (at each moment) the goal that is currently more important to her may result in a balancing of multiple goals, but by hypothesis, there can be no question of whether the balance achieved is rational relative to A's enduring assessment because there is no such assessment.

I will leave open the question of whether instrumental rationality requires not only that one serve one's goals well and in accordance with one's assessment of their relative importance but also that one's goals and one's judgments regarding their relative importance remain fairly stable over time. Since I will not be focusing on stretches of time that include significant ranking reversals, the question of whether there is some connection between being instrumentally rational and having fairly stable evaluative judgments will not arise, and my reasoning will not depend on any such connection.[10]

---

[8] Of course, if A has additional options and A's preferences are, say, intransitive, things will be more complicated. Since these complications are tangential relative to my aim in this chapter, I will put them aside.

[9] The broad brushstrokes defense of the suggestion that I will offer can, I assume, be developed and refined in light of the work of artificial intelligence (AI) researchers, operating system designers, and psychologists pursuing related inquiries. My aim at this point is just to say enough relative to my purpose of developing a novel philosophical line of thought about temptation.

[10] For a sense of some of the issues in the current debate regarding whether there might be some such connection, see, for example, Bratman (2012). For some influential contemporary discussion regarding temptation in which ranking reversals figure crucially, see, for example, Holton (2009)

IV. I turn now to the following question. How might a motivational system that helps an agent efficiently balance multiple ongoing goals also leave the agent susceptible to temptation? Toward answering this question, I sketch out a simple scenario on which my response will build.

Consider the following case involving an agent A and three kinds of edible goods, X, Y, and Z. A's three ongoing goals are consume X, consume Y, and consume Z. A must pursue X, Y, and Z one at a time, and (in this case) none is (judged) more important than the other.

Suppose that A has a simple "turn-taking" motivational system in which (1) if A is not currently consuming anything, A is most motivated to pursue consuming whichever of X, Y, and Z he has not consumed in the longest time, and (2) once A is consuming something, he is highly motivated to continue pursuing the consumption of that thing for a certain stretch of time but then loses this motivation, which is replaced by the motivation to pursue what he has consumed longest ago. A big advantage of this system is that A is not constantly pulled in different directions. A big disadvantage of the system is that A's pursuit of his goals is too rigidly structured to allow him to take advantage of "affordances" that might present themselves as he navigates through his environment (where an *affordance* is to be understood as an opportunity that is readily afforded by one's current circumstances). Suppose, for example, that X, Y, and Z are not always ready at hand (or feasibly storable for later consumption), and A has stumbled on a bumper crop of X. A might be better served by loading up on X than by pursuing the consumption of X for only the usual stretch of time.

Of course, a system in which A is very heavily guided by affordances can have its own problems. If, for example, A's motivation to pursue something is spurred when its pursuit is primed by, for example, its being present and dampened when it is not, A may end up neglecting even a very important goal, since his motivation to pursue the goal may be dominated by the presence of goal-related objects rather than by its importance. In a fairly stable but not perfectly predictable world, A might be well served by a hybrid motivational system that is influenced by affordances (among other things) but also exerts some turn-taking pressure.

Still, given the possibility of dramatic unexpected changes, a system that has been working well might, if left completely unmonitored, suddenly fail. If, for example, the availability of, say, Z skyrockets and Z continues to be available at record-high levels (so that affordances with respect to

and, relatedly, Ainslie (2001). Johanna Thoma's chapter in this volume (Chapter 1) makes interesting contributions to both debates.

consuming Z become extremely common), a system that worked well when Z remained within its prior ranges might prompt A to overconsume Z (relative to its importance).

If A is capable of detecting this mismatch, he may be able to cope with his problem by engineering an intervention that compensates for the overavailability of Z. For example, he might reduce his navigation on routes along which Z is present. Or he might introduce competing primes for consuming X or Y instead. He might, for instance, form an implementation intention to pursue consuming Y whenever (cue) C is present and so be primed by the presence of C to pursue consuming Y.[11]

V.   The hybrid motivational system described in the preceding section (which has both turn-taking and affordance-priming features) and the temptation-resisting corrective measures mentioned at the end of the section are not just theoretically possible. There is evidence that turn-taking features and susceptibility to priming (of goal-relevant action by the presence of goal-relevant objects) characterize the human motivational system, as well as evidence that humans can and do employ, with substantial (but certainly not perfect) success, the corrective measures of prime avoidance and construction.[12]

Recognizing this can shed a great deal of light on our susceptibility to temptation and on the nature of (instrumental) rationality. Notice first that insofar as an agent's motivational system includes turn-taking features and susceptibility to priming, the strength of the agent's motivation to pursue a goal (and engage in goal-directed action) will vary even if her ranking of the goal remains fixed. And this makes room for the possibility of temptation and of succumbing to temptation, where such succumbing involves a pattern of motivation and action that does not line up with the agent's ranking of her goals. If, for example, one of an agent's ongoing goals includes (occasionally) enjoying sweet treats, and the agent's motivational system includes susceptibility to priming, then if sweet treats become very readily available, the agent's frequently spurred motivation to consume them can easily prompt a pattern of consumption that she judges to be

---

[11]   Success in relation to general or vague goals is often facilitated by the formation of more specific implementation intentions, which detail the when, where, and/or how of goal-directed behavior (Brandstätter, Lengfelder, and Gollwitzer 2001). If, for example, I have the goal of consuming fish oil regularly, I can (other things being equal) increase my chances of actually doing so by forming a more specific implementation intention, such as the intention to make fish for dinner every Friday (after stopping by the fish market on my way home from work) or the intention to take a fish oil supplement every morning as soon as I finish breakfast.

[12]   Much of the relevant psychological literature is helpfully reviewed in Huang and Bargh (2014).

unacceptable. This, in turn, leaves plenty of room for instrumental rationality to play a monitoring and corrective role, with intentions figuring as an important tool for managing and creating primes.

Notice that when an agent is overconsuming sweets and seeks to correct for this, it might appear as though she is struggling against an "alien" (or disowned) desire (to enjoy sweets) or else against the purportedly unadulterated curse of a motivational system that allows for a mismatch between the weight or ranking of a goal and its motivational force. In reality, however, the agent may value enjoying sweets, benefit from the motivational strength of her goals varying even if the ranking of her goals remains fixed, and be struggling due to contextual factors that result in the excessive priming of her (endorsed) goal (of enjoying sweets). The agent may seek to cope with her problem via the bright-line personal rule of "no sweets at all," but the rigid, single-minded favoring of nutritious food may be far from ideal, even if the agent manages to stick to her rule and avoid lapses that compromise her sense of self-control.[13] A more nuanced rule or habit may be both feasible and superior.

It is worth emphasizing that I am not denying that there may be a form of temptation that involves being drawn to something that one assesses as invariably unworthy of pursuit and that this form of temptation has its own, perhaps more uniformly lamentable, dynamic.[14] Whatever one's view about whether this form of temptation exists or can be excised, one can, I take it, still agree with the suggestion that the sort of susceptibility to temptation I have been exploring is, for us, among the "certain things in life."

It might be suggested that an ideally rational agent could avoid temptation of the sort I have been exploring by never being directly motivated by her goals but motivated only by a comprehensive master plan or policy that endorses the "right" mix of turn taking and affordance-spurred motivation for every possible world. (Relatedly, it might be suggested that an ideally rational agent would fill in and refine her goals so that there is no room for any tension between them.) Even if this is correct, such policies (and related hyper-filled-in-and-refined goals) are too cognitively expensive to be feasible. Maximally specific plans, which include guidance for every possible moment given every possible contingency, are not a genuine option; our plans must be filled in as we go along, with refinements or

---

[13] For an extensive and illuminating discussion of the "dangers of willpower," including both worries regarding rigidity and the contrived magnification of the significance of lapses, see Ainslie (1999).
[14] See, relatedly, note 6.

adjustments made as complications or "entanglements" arise or are anticipated.[15] We are, it seems, stuck with a (resistible) susceptibility to temptation and not as a gratuitous anomaly but as a challenge that goes hand-in-hand with the features of our motivational system that allow us to balance multiple ongoing goals.

To summarize, turn-taking and affordance-priming features make for a motivational system with potentially efficient balancing power. They also, however, make for a motivational system in which some susceptibility to temptation is to be expected. Importantly, primes are not always just found; they can be created (via, for example, implementation intentions), and created primes can, at least sometimes and to some extent, correct for the susceptibility to temptation. (Of course, primes can also be created by individuals keen on [or at least not wary of] exacerbating our susceptibility to temptation. And when the primes are integrated without warning and with the assistance of personal information, such as, for example, with internet ads or offers prompted by an individual's internet browsing patterns, they can be particularly powerful.)

VI. In connecting our susceptibility to temptation to a type of motivational system that is generally quite well suited to the task of balancing multiple ongoing goals, I do not mean to suggest that cases in which this susceptibility leads an agent astray are not really cases of irrationality. Although, in my view, failures of this sort are understandable and we should, indeed, count on them sometimes occurring, they can, I allow, still be properly counted as failures of instrumental rationality and of rational self-control. The agent can and should recognize that her choices are adding up in a way that is clearly inadequate for success by her own lights and correct for the mismatch between her assessment of the importance of her goal and its motivating force.

Of course, insofar as the criterion for success is somewhat vague, whether or not an agent is giving into temptation can be correspondingly fuzzy (even assuming, as we have been, that the agent's goals and her assessment of their relative importance remain unchanged). There is, for example, normally some gray area when it comes to how many treats one can consume per month before one is failing to have no more than the occasional treat. Relatedly, it may be that an agent's sense that her behavior

---

[15] For a discussion of some of the ways in which our plans are normally partial, see Bratman (1983: sec. 2). In the same section, Bratman discusses the role of deliberation and plans in intelligent activity, quoting John Dewey on deliberation's "beginning in troubled activity" characterized by "entanglements" (Dewey 1957: 199).

qualifies as giving in to temptation only gradually solidifies so that the conviction that she has failed herself is neither clearly anticipated nor catches her completely by surprise. Arguably, this occurs in at least some cases of procrastination in which the agent has a vague goal that is never supplemented by an implementation intention. The following variation of a case presented by Alison McIntyre seems like it could easily fit the bill: a professor carries around a set of student papers with the goal of grading them very soon, never forms a specific implementation intention, experiences a growing feeling of dissonance, and ultimately scolds himself for having procrastinated – having graded the papers weeks later, the night before final grades are due.[16] If, by contrast, an agent forms a clear implementation intention and judges (and persists in judging) that he should act accordingly and yet fails to do so, his sense that he is giving in to temptation will be more cut and dry. Consider, to pick up on an example Alfred R. Mele discusses in this volume (Chapter 9), the case of someone who is watching TV but has the goal of getting back to work "very soon," and who forms the implementation intention to get back to work when the next commercial break starts, judges that it is best to act accordingly and yet fails to do so even though his judgment remains fixed.[17] This agent's sense of giving into temptation will, presumably, arise with full force all at once. In the end, whether the dissonance experienced by the procrastinating agent in each of the two cases under consideration arises gradually or all at once, it fits neatly with the idea that a failure of instrumental rationality is indeed in play.

VII. I turn now to a discussion of how the motivational system I have been focusing on makes sense of many prominent philosophical views about temptation, including the following: (1) there is room for cases of temptation that are not cases in which the agent's ranking of a tempting option changes as the tempting opportunity approaches and looms large, (2) although temptation is often heightened by visceral factors, it is possible to be tempted to overengage in activities that give no pleasure, and (3) temptation can threaten even if ultimate goals are not criticizable. I will touch on each of these views in turn.

There has been increasing discussion, in the philosophical literature on temptation, of cases involving a ranking reversal.[18] In paradigmatic cases of

---

[16] See McIntyre (2006). For an interesting follow-up discussion of McIntyre's case, see Stroud (2010).

[17] Strictly speaking, this case is a variation on Mele's example because, in Mele's example, the agent first judges that it would be best to get back to work when the next commercial break starts and then resolves to do so.

[18] See Ainslie (2001), Bratman (1999), and Holton (2009).

the relevant sort, the agent ranks V-ing at time t above W-ing at time t, until the time for action is imminent or at hand, at which point his ranking temporarily reverses and he ranks W-ing at time t above V-ing at time t. Consider, for example, the following case: an agent prospectively ranks studying for an exam above going to an upcoming party, but when the night of the party arrives, her ranking reverses, and she ranks going to the party above studying for the exam; the next day, she regrets not having stayed home and studied. Now it may well be true that, in at least some cases of this sort, W-ing at time t amounts to succumbing to temptation. Still, it is often conceded that there is room for cases of giving in to temptation that do not involve this sort of reversal and which therefore cannot be explained via mechanisms that generate such reversals. (Significantly, the possibility of such cases, and, in particular, of giving into temptation without having been lured into reconsidering one's priorities by some change in perspective, has also often seemed puzzling.[19]) The picture of temptation that I have painted (which may not exhaust the forms the phenomenon can take) requires that the tempted agent's motivation vary over time but does not require a ranking reversal. It is thus well suited to accommodate cases of giving in to temptation in which the agent does not reconsider her priorities. Suppose, for example, that one believes that one should eat sweets only occasionally. Even if how much one values eating a sweet does not increase as the moment it can be consumed approaches, if sweets suddenly become much more pervasive, a susceptibility to priming will result in a frequently spurred motivation to eat sweets, with resistance requiring some sort of corrective intervention.

This relates to the idea that temptation is often heightened by visceral factors. Because visceral factors provide palpable, hard-to-miss "suggestions" for action, a susceptibility to priming can lead to overconsumption when visceral factors proliferate and are not easily avoided or countered. If, to continue with the preceding example, sweets are not only more readily available but also visually present, priming of the goal to consume sweets is heightened. But, since priming is possible even for ongoing goals that give no pleasure, as might, for example, be the case with the goal of checking a certain email account or messaging system, constant cues priming a goal can result in excessive engagement in activity directed at that goal (relative to how much the goal is judged to matter) even if the goal gives no pleasure. One thus can find it difficult to refrain from engaging in an activity that one does not find pleasant or important. Where implementation

---

[19] For an influential specification of the puzzle, see Davidson (2001 [1970]).

intentions seem futile, temptation may shade into compulsion, even in the absence of any "alien" desires.

Because the picture of temptation that I have painted allows that a pattern of motivation can be criticizable relative to the agent's assessment of the relative importance of her goals, it makes sense of the idea that temptation can threaten even if ultimate goals are not criticizable. If, by contrast, it is assumed that an agent's goals are to be equated with her motivations and that her motivations are uncriticizable, we must conclude that there is no basis for concern about facing or succumbing to temptation; for, if motivations are uncriticizable, then, as Hume famously insists, acting on one's strongest motivations cannot involve a rational failure (as long as one's motivations are not based on false beliefs), even if a "trivial good" produces a disproportionately strong desire due simply to the "advantage of its situation" (such as, for example, its visibility or close proximity).[20] My position conflicts with Hume's, but it does not go to the opposite extreme and endorse the assumption that if a trivial good produces a disproportionately strong desire due simply to the advantage of its situation, acting accordingly must involve a rational failure. For, according to my position, even when things are going well, the balancing of goals involves allowing motivational strength to vary so that one is not moved to action only by one's most important goal(s).

VIII.   I have argued that, in familiar cases such as the one I opened with, susceptibility to temptation is not an anomaly but is instead integral to a type of motivational system that is generally quite well suited to the task of balancing multiple ongoing goals. The type of temptation that I have focused on is consistent with an instrumental conception of practical rationality: giving in to temptation can figure as a rational failure even if the agent's goals and his assessment of them provide the only standards of rationality. Moreover, susceptibility to temptation is possible even in the absence of a ranking reversal, "alien" desires, or an unadulterated curse involving gratuitous motivational variability. Indeed, it is, even apart from such complications, to be expected.

References

Ainslie, G. 1999. The dangers of willpower. In *Getting Hooked*, ed. J. Elster and O.-J. Skog. Cambridge: Cambridge University Press.

---

[20] Hume 1978 [1739–40]: 416.

Ainslie, G. 2001. *Breakdown of Will.* Cambridge: Cambridge University Press.

Brandstätter, V., A. Lengfelder, and P. Gollwitzer 2001. Implementation intentions and efficient action initiation. *Journal of Personality and Social Psychology* 81:946–60.

Bratman, M. 1983. Taking plans seriously. *Social Theory and Practice* 9:271–87.

Bratman, M. 1999. Toxin, temptation, and the stability of intention. In *Faces of Intention.* Cambridge: Cambridge University Press.

Bratman, M. 2007. A desire of one's own. In *Structures of Agency.* Oxford: Oxford University Press.

Bratman, M. 2012. Time, rationality, and self-governance. *Philosophical Issues* 22:73–88.

Bratman, M. 2014. Temptation and the agent's standpoint. *Inquiry* 57:293–310.

Davidson, D. 2001 [1970]. How is weakness of the will possible? In *Essays on Actions and Events.* Oxford: Oxford University Press.

Dewey, J. 1957. *Human Nature and Conduct.* New York, NY: Random House.

Frankfurt, H. 1971. Freedom of the will and the concept of a person. *Journal of Philosophy* 68:5–20.

Holton, R. 2009. *Willing, Wanting, Waiting.* Oxford: Clarendon Press.

Huang, J. Y., and J. A. Bargh. 2014. The selfish goal. *Behavioral and Brain Sciences* 37:121–35.

Hume, D. 1978 [1739–40]. *A Treatise of Human Nature*, ed. L. A. Selby-Bigge and P. H. Nidditch (2nd edn). Oxford: Clarendon Press.

McIntyre, A. 2006. What is wrong with weakness of will? *Journal of Philosophy* 103:284–311.

Mele, A. In press. Exercising self-control: an apparent problem resolved. In *Self-Control, Rationality, and Decision Theory*, ed. J. L. Bermúdez. Cambridge: Cambridge University Press.

Nuyen, A. T. 1997. The nature of temptation. *Southern Journal of Philosophy* 35:91–103.

Stroud, S. 2010. Is procrastination weakness of will? In *The Thief of Time: Philosophical Essays on Procrastination*, ed. C. Andreou and M. D. White. Oxford: Oxford University Press.

Thoma, J. In press. Temptation and preference-based instrumental rationality. In *Self-Control, Rationality, and Decision Theory*, ed. J. L. Bermúdez. Cambridge: Cambridge University Press.

Watson, G. 1975. Free agency. *Journal of Philosophy* 72:205–20.

Watson, G. 1987. Free action and free will, *Mind* 96:145–72.

CHAPTER 8

Frames, Rationality, and Self-Control

*José Luis Bermúdez**

Being influenced by how outcomes are framed is widely held to be a paradigm of irrationality. Yet experimental and theoretical discussions of framing have ignored the most interesting situations in which framing occurs. These are when an agent or decision maker consciously and simultaneously considers a single outcome or scenario under two different frames and evaluates it differently in each frame. From the perspective of standard theories of rational decision making and rational choice, such *clashes of frames* are not supposed to happen – and if they do occur, they are paradigms of irrationality. The standard view, I claim, gets things almost completely wrong (Bermúdez forthcoming). There is nothing intrinsically irrational about seeing the world in conflicting frames. In fact, actively engaging competing frames can be a powerful force for rationally engaging with problems that classical theories of rationality are unable to tackle.

This chapter applies this basic idea to self-control. Clashes between temptation and self-control can be modeled as clashes of frames, and, I argue, successful self-control can often be a matter of how one frames oneself and one's goals. Agents exercising self-control can behave perfectly rationally despite (and in fact because) they are consciously and simultaneously considering a single outcome or scenario under two different frames and evaluating it differently in each frame.

8.1   Framing Effects: The Orthodox View

The concept of framing is extremely important in the psychology of reasoning and has become a cornerstone of behavioral economics and behavioral finance. It has been experimentally studied in many different ways – in laboratory experiments, via large-scale data mining of investment

* Work on this chapter was supported by a grant from the Philosophy and Psychology of Self-Control Project, funded by the John Templeton Foundation.

decisions, and through functional neuroimaging. The consensus view is that susceptibility to framing effects is a paradigm of irrationality.

The extensive experimental literature on framing effects goes back several decades. Amos Tversky and Daniel Kahneman were responsible for some of the most significant early studies, including the celebrated Asian disease paradigm (as presented, for example, in Tversky and Kahneman 1981). These studies have been taken (by their authors and many others) to reveal a significant gap between the messy realities of practical reasoning and the idealized norms of Bayesian decision theory. When Tversky and Kahneman developed *prospect theory*, their model of how they think that practical decision making actually works as opposed to how normative decision theory thinks that it ought to work, they placed framing at its core. All practical reasoning, they think, starts with an editing phase that, in effect, imposes a frame on the decision problem. Their motivation for developing prospect theory comes out in the following passage:

> Because framing effects and the associated failures of invariance are ubiquitous, no adequate descriptive theory can ignore these phenomena. On the other hand, because invariance (or extensionality) is normatively indispensable, no adequate prescriptive theory should permit its violation. Consequently, the dream of constructing a theory that is acceptable both descriptively and normatively appears unrealizable.[1]

So, because framing is both irrational and unavoidable, no theory of reasoning can be both descriptively adequate and normatively acceptable.

Framing effects have been revealed in many different contexts.[2] To get the flavor of the experiments, here are two figures that offer a schematic representation of the family of experiments standardly taken to show that people are risk averse for gains but risk seeking for losses. Figure 8.1 illustrates the basic design.

Subjects are presented with a choice between a certain outcome and a risky outcome. For one group of subjects, the choice is framed positively (in terms of the number of lives saved, for example), whereas for a second group, it is framed negatively (in terms of the number of lives lost).[3]

---

[1] Tversky and Kahneman 1981: S272. [2] For a survey, see Bermúdez (forthcoming).
[3] In the original experiment, subjects were asked to imagine that the country was preparing for the outbreak of a disease originating in Asia and expected to kill 600 people. They were asked to choose between two programs with predictable outcomes. In the positive frame, the outcome of first program was described as saving 200 lives out of 600, while the outcome of the second program was described as a 1/3 probability of saving all 600 people and a 2/3 probability of saving none. In the

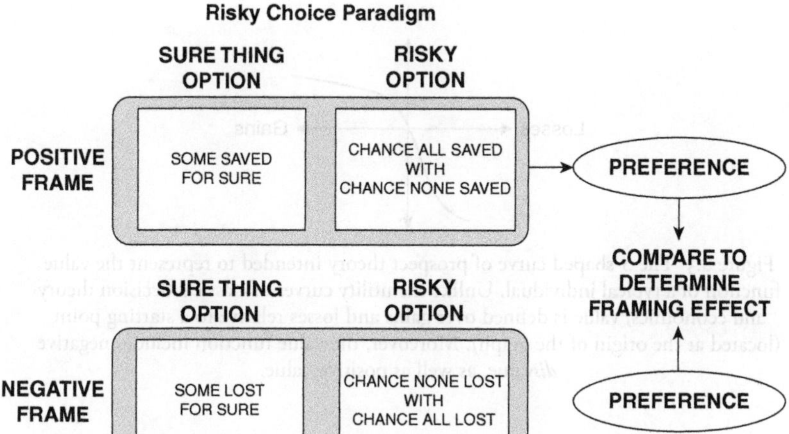

Risky Choice Paradigm

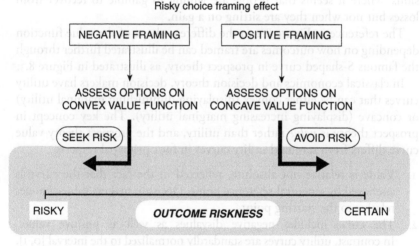

Figure 8.1 The standard risky-choice framing paradigm

Figure 8.2 The effect of framing on attitudes toward risk

Tverksy and Kahneman discovered a preference reversal across the two frames, as illustrated in Figure 8.2.

Subjects in the positive frame show a strong preference for the certain outcome, while subjects in the negative frame prefer the risky outcome.

negative frame, the outcomes were described as 400 people dying, on the one hand, and a 1/3 probability of nobody dying and a 2/3 probability of all 600 dying, on the other.

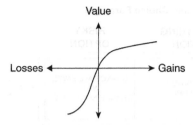

Figure 8.3 The S-shaped curve of prospect theory intended to represent the value
function of a typical individual. Unlike the utility curves of classical decision theory
and economics, value is defined over gains and losses relative to a starting point
(located at the origin of the graph). Moreover, the value function includes negative
*dis*value, as well as positive value.

So the conclusion is drawn, people are risk averse for outcomes that are
framed as gains and risk seeking for outcomes that are framed as losses.[4]
This basic pattern has been revealed in a wide range of investment decisions, where it seems that people are willing to gamble to recover from
losses but not when they are sitting on a gain.

The reference in Figure 8.2 to the different shape of the value function
depending on how outcomes are framed can be illustrated further through
the famous S-shaped curve in prospect theory, as illustrated in Figure 8.3.

In classical economics and decision theory, decision makers have utility
curves that can be either convex (displaying diminishing marginal utility)
or concave (displaying increasing marginal utility). The key concept in
prospect theory is *value* rather than utility, and the prospect theory value
curve differs from standard utility curves in four principal respects.

1. Value is relative not absolute, reflected in the fact that the curve is
    anchored by a neutral reference point. Decision makers value changes
    relative to the starting point.
2. The curve includes negative disvalues as well as positive values.
    In contrast, utility curves are standardly normalized to the interval [0, 1].
3. The property of *loss aversion* is built in. Losses seem to be more
    psychologically impactful than gains, and hence the slope is steeper
    for losses than for gains.
4. The curve displays both diminishing marginal value and diminishing
    marginal disvalue.

---

[4] The original Asian disease experiment was across subjects, but it has been replicated within subjects
(i.e., with individual subjects displaying the preference reversal). See Kühberger (1995), for example.

The S-shaped value curve captures (as intended) the principal psycho-logical tendencies revealed in the classical framing effect experiments. Relative to the reference point, negative frames engage risk-seeking beha-vior, while positive frames encourage risk aversion.

As a psychological theory, prospect theory has no claim to normative validity. But, as illustrated in the passage quoted earlier from Tversky and Kanhemann, it is predicated on a normative ideal of completely frame-independent reasoning. This normative ideal is eloquently articulated in the following passage from Kenneth Arrow:

> A fundamental element of rationality, so elementary that we hardly notice it, is, in logicians' language, its extensionality. The chosen element depends on the opportunity set from which the choice is to be made, independently of how that set is described. To take a familiar example, consider the consumer's budget set. It is defined by prices and income. Suppose income and all prices were doubled. Clearly, the set of commodity bundles available for purchase is unchanged. Economists confidently use that fact to argue that the chosen bundle is unchanged, so that consumer demand functions are homogeneous of degree zero in prices and income. But the description of the budget set, in terms of prices and income, has altered. It is an axiom that the change in description leaves the decision unaltered.[5]

This claim about the normativity of frame-independent reasoning will be the focus of the remainder of this chapter. I begin in the next section with some general reasons for thinking that frame sensitivity can be rational. The remainder of the chapter focuses on frame sensitivity in the specific context of self-control.

8.2   Framing Effects and Rationality

There is almost complete consensus that susceptibility to framing effects is irrational – indeed, not just irrational, but a paradigm of irrationality. Explicit arguments are rarely given, however. My best interpretation of the quotation from Arrow is that extensionality follows from the fact that normative theories of rationality are typically formulated set theoretically. For Savage, for example, events are sets of states. And sets are certainly extensional. But this would be an odd way to derive a basic principle of rationality.

Another possibility would be some form of money-pump argument. An agent who is susceptible to framing effects will assign different utilities to a single outcome a depending on how it is framed. For example, he might assign greater utility to $f_1(a)$ than to $f_2(a)$. So, if b is some outcome such that $u[f_1(a)] > u(b) > u[f_2(a)]$, then presumably the agent will be prepared to pay some sum of money, say ε, to switch from outcome $f_2(a)$ to outcome b and then some additional sum of money, say σ, to switch from outcome b to outcome $f_1(a)$. But then he is back where he started and poorer by $\varepsilon + \sigma$. This argument is not very compelling, however. If the agent is unaware that $f_1(a) = f_2(a)$, then the issue seems to be ignorance rather than irrationality. But, if the agent is aware that $f_1(a) = f_2(a)$, then there seems to be no reason why he should be prepared to pay to make the switches.

Still, one might reasonably wonder (from a normative perspective) whether the contrast between awareness and unawareness should be drawn so starkly. What about the agent who is unaware that $f_1(a) = f_2(a)$ but who really should be aware that these are really two different perspectives on a single outcome? Something like this seems to be going on in the classical experiments. One index is that subjects typically reverse their judgments when the framing effect is revealed. In this respect, experimental framing effects are unlike the Allais and Ellsberg scenarios, where many subjects stick with their choices even when they see how those choices contravene basic principles of utility theory.

I will step back from this issue, however, in order to focus instead on cases of frame-sensitivity and frame-dependent reasoning where the issue of judgment reversal simply does not arise. These are cases where subjects are well aware that there is a single outcome framed in two (or more) different ways and that the value they assign to that outcome varies by frame. It may be helpful to anchor the discussion in a more familiar philosophical context. Contexts that permit departures from extensionality are typically termed *intensional contexts*. The defining feature of an intensional context is that it allows for (truth- and/or rationality-preserving) failure of substitution of materially equivalent sentences. By extension, a *hyperintensional context* allows for (truth- and/or rationality-preserving) failure of substitution of logically equivalent sentences.

Intensional and hyperintensional contexts have been well studied. Not so what I term *ultraintensional contexts*. An ultraintensional context is a context that allows for (truth- and/or rationality-preserving) failure of substitution of *known* identities. Belief contexts are quite plainly not ultraintensional. It is rational for me to believe that Tintoretto was known as Il Furioso while not believing that Jacopo Comin was known as Il Furioso – but only if I am unaware that Tintoretto was born Jacopo Comin.

There certainly are ultraintensional contexts if we stretch the definition to include known logical equivalences as well as known identities. Consider, for example, the complex predicate "____ is the Gödel number of a proof of ____," and let T_1 and T_2 be two provable mathematical theorems that are provably equivalent and known to be so. Then, if "*n* is the Gödel number of a proof of T_1" is true, "*n* is the Gödel number of a proof of T_2" is certain to be false.

In my book *The Power of Frames*, I use the example of framing effects to argue that valuation is an ultraintensional context. It *can* be perfectly rational, I argue, to assign different values to a single outcome under different frames, even when the decision maker is completely aware that there is a single outcome at stake. In other words, evaluative contexts can permit rationality-preserving failure of substitution of *known identities*. Rationality is not *always* preserved, but at least sometimes it is. Evaluative contexts can present many examples of irrational frame dependence, but frame-dependent reasoning can be rational. In fact, it can also solve problems that have remained intractable to orthodox theories of rational decision making. This takes us to the main topic for this chapter, which is self-control.

8.3 Self-Control: The Problem with Resoluteness and Resolution

Self-control can be exercised in many contexts, but for present purposes, I want to restrict attention to a relatively straightforward example – what might be termed the *paradigm case*. The paradigm case illustrates some of the basic issues raised by self-control in a very clear form, and it seems likely that a satisfying account of the paradigm case will scale up to more complex examples.

The paradigm case of self-control is an act (or omission) performed at a specific moment in time (the moment of resisting temptation). Two factors make that act (or omission) an exercise of self-control. The first is that it in some sense implements an earlier decision, plan, or commitment. The second is that, at the moment of choice, current desires motivationally outweigh the earlier decision, plan, or commitment. So the paradigm case of self-control involves a clash at the moment of choice between the motivational force of a long-term desire for a future benefit and the motivational force of a short-term desire for an immediate benefit.

Figure 8.4 is a familiar diagrammatic representation of the paradigm case, illustrating one popular mechanism for explaining how the motivational force of a long-term desire for a future benefit can come to be

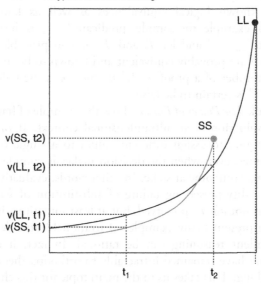

Figure 8.4 Illustration of how hyperbolic discounting can necessitate the exercise of
self-control at time t_2. SS is an immediate benefit (smaller, sooner), while LL is
a long-term benefit (larger, later).

outweighed by the motivational force of a short-term desire for an immediate benefit. In the figure, "SS" stands for "smaller, sooner" – i.e., the short-
term reward – while "LL" stands for "larger, later" – i.e., the long-term
reward.

Figure 8.4 illustrates the process of hyperbolic discounting.
In hyperbolic discounting, the ratio between the degree of discounting at
the beginning and end of a given temporal interval varies as a function of
the length of time to the benefit. The ratio is smaller when the delay is long
but increases as the benefit gets closer.[6] So, as the moment of choice
approaches, the hyperbolic discount function for SS steepens and the
value assigned to SS surpasses the value assigned to LL.

In the paradigm case, self-control is exercised at the moment of choice
just if the decision maker resists SS and sticks with the plan to delay

---

[6] The contrast is with exponential discounting, where the ratio of change per unit delay is constant, so
the value attached to a future reward will always increase the same amount from day to day,
irrespective of whether the reward is expected tomorrow or next year.

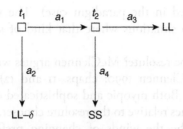

Figure 8.5 The paradigm case of self-control represented as a sequential choice problem. The moment of planning is at time t_1 with the moment of choice at time t_2. At t_1 the agent has a choice between making a precommitment to LL (which would guarantee receiving LL, but at a cost, namely δ) or continuing to t_2. At t_2 the choice is between SS and LL.

satisfaction until LL is realized. This can be represented as a sequential choice problem, illustrated in Figure 8.5

Time t_1 is the moment of planning and time t_2 the moment of choice. At time t_1 the agent prefers LL to SS and so plans to continue across to the moment of choice at t_2. Between t_1 and t_2 the agent has a preference reversal and comes to find SS more attractive than LL. So, at t_2 she has to decide whether to follow through on her earlier commitment to LL or whether to follow her current preference for SS. At time t_1, however, the agent has a further option. She can make a precommitment to LL to avoid problems created by the anticipated preference reversal. The precommitment strategy would guarantee her not succumbing to SS, but at a price, namely δ. So the payoff from precommitment is LL–δ.

There is a well-established terminology for describing the different options in Figure 8.5 (comprehensively discussed in McClennen 1990, for example). An agent who goes across at time t_1 but then is overwhelmed by temptation at time t_2 (i.e., who takes a_1 followed by a_4) is said to be a *myopic chooser*. The myopic chooser simply acts according to her preferences at the moment of choice, neither looking ahead to future preferences and future choice nodes nor looking back to earlier plans and commitments. In contrast, a *sophisticated chooser* will look ahead and take into account what will and will not be feasible at future choice nodes. A sophisticated chooser who anticipates a preference reversal will see that LL will not be a feasible outcome at t_2 and so will opt for some form of precommitment strategy (e.g., a_2). Finally, a *resolute chooser* will stick to her plan at t_2 and continue across to LL, disregarding her temporary preference for SS over LL. Only the resolute chooser displays self-control (at least in the sense envisioned in the paradigm case). The sophisticated chooser avoids getting into situations where that kind of self-control would be necessary.

Is it rational to be resolute? McClennen argues with considerable force that it is (see McClennen 1990: chaps. 11 and 12). His arguments are pragmatic in form. Both myopic and sophisticated choosers end up with suboptimal outcomes relative to the resolute chooser. The myopic chooser is blown around by the winds of changing preferences, whereas the sophisticated chooser is forced to pay the price of precommitment. In effect, McClennen argues that the rational agent will sail a steady course to the planned destination while cutting out the middleman.

If resolute choice can be rational, then there can be rational failures of what is often termed the *separability principle*. The separability principle says that the decision problem an agent confronts at a particular choice node is completely determined by her preferences over the available options and payoffs at that choice node. So, at time t_2 in the paradigm case, the agent is simply choosing between a_3 and a_4 and should take into account only her preference at that time for SS over LL. The separability principle governs both myopic and sophisticated choice – obviously in the case of the myopic chooser and in the case of the sophisticated chooser because his reasoning about what strategies will be feasible in the future assumes separability.

Quite plainly, though, saying that it can be rational to be resolute, and hence for the separability principle to fail, leaves a very important problem unsolved. The supporter of resolute choice still has to explain how resoluteness is possible. At time t_2 the agent prefers SS to LL. Lifting the requirement of separability tells us that the agent's practical reasoning at t_2 can extend beyond her preferences at t_2, but as yet we have no model of what such reasoning might look like or of how factors not directly in play at time t_2 might feature in it.

McClennen himself focused primarily on the decision-theoretic and normative dimensions of resolute choice, leaving open the question of the actual mechanisms by which resolute choice might be achieved. Richard Holton has made an important contribution to filling this gap. As part of a more general discussion of intention and the will, Holton develops an account of strength of will that he suggests "might be thought of as providing philosophical underpinnings for McClennen's work" (Holton 2009: 141).

The centerpiece of Holton's account is the notion of a resolution. *Resolutions* are motivating states (what Davidson termed *proattitudes*), but they are not desires. Whereas traditional, Humean accounts of action identify the springs of action with beliefs and desires, Holton argues that the process of deliberation typically culminates in an intention to act in a certain way. *Intentions* are irreducible, psychological states. They often entail beliefs about how one will behave, but they are not reducible to such beliefs (or to any other beliefs), and while they can be caused by desires, they can persist even when the desire that generated them has disappeared or radically changed.

Resolutions, for Holton, are a particular species of intention. They are intentions formed for a specific purpose:

> Resolutions are contrary inclination defeating intentions: intentions formed by the agent with the very role of defeating any contrary inclinations that might emerge. (Holton 2009: 119)

A resolution, then, is an intention with a specific eye, as it were, on contrary inclinations and other temptations. Resolutions are formed with the full knowledge that resistance to temptation will most likely be required. But how does this resistance work in practice? According to Holton, it involves the exercise of willpower. One of the mechanisms of willpower is what Holton calls "rehearsal." To rehearse a resolution is to remind oneself of it, and to remind oneself also of the reasons for which one is holding it. Rehearsing a resolution contrasts with reconsidering it, which robs it of its efficacy. One cannot reconsider a resolution, he suggests, without suspending it. In fact, the process of rehearsal works as an antidote to reconsideration.

One beneficial side effect of Holton's account is that it points toward an account of when it is rational to be resolute (see Holton 2009: chap. 7). So, for example, it is rational *not* to reconsider a resolution when confronted with the very temptations against which the resolution is intended to guard or when one might reasonably think that one's judgment will be impaired in ways that it was when the resolution was formed. Reconsideration may be rational, in contrast, when circumstances have changed or when one thinks that the reasoning that led to the resolution might have been flawed.

So, here is my interpretation of how Holton's account would work in the paradigm case. The decision maker forms her resolution at time t_1, in full knowledge of the preference reversal likely to take place between t_1 and t_2. Unlike the sophisticated chooser, however, she is confident of her ability to resist contrary inclinations and so eschews a_2, the available precommitment strategy, instead taking a_1 across to the second choice node. At that choice node, she resists the siren call of SS by rehearsing her earlier resolution.

In doing this she breaks the hold of the separability principle and does so perfectly rationally because she knows that SS is precisely the temptation that her resolution was intended to ward off. Guided by her earlier resolution she takes strategy a_3 and reaps the reward of LL.

It is a very neat account. Still, despite its neatness, it seems to me to confront three significant problems. The first has to do with how Holton characterizes the process of succumbing to temptation. The second involves whether his account can ultimately be differentiated from the sophisticated choice strategy. Third, Holton leaves unexplained what it would be for the process of reconsidering a resolution to lead not to its rejection but rather to its reaffirmation.

For Holton, failing to abide by a commitment seems to be a very reflective process. The core of his account is that to succumb to temptation is, in effect, to reconsider a resolution. Here is how he describes what goes on in reconsidering a resolution:

> I suggest that the full-blown reconsideration of a resolution does involve *suspension* of that resolution. To fully reconsider a resolution is to open oneself to the possibility of revising it if the considerations come out a certain way; and that is to withdraw one's current commitment to it. Someone might say that the resolution remains in place pending the outcome of the revision. But such a claim is not very plausible. For much of the point of a resolution, as with any intention, is that it is a fixed point around which other actions – one's own and those of others – can be coordinated. To reconsider an intention is exactly to remove that status from it. (Holton 2009: 122–23)

This seems on the face of it to overintellectualize a very common phenomenon.[7] Consider the sequential choice version of the paradigm case. What happens between time t_1 and time t_2 is simply that the agent undergoes a preference reversal. The attractiveness of the long-term goal of *mens sana in corpore sano* recedes in the face of the imminent appearance of, say, a slice of chocolate cake. It is true that the commitment at time t_1 to hold on for LL loses its force as SS becomes more attractive. But a commitment does not have to be rationally reconsidered to start seeming less compelling as the moment of truth approaches. In fact, the hyperbolic discounting model illustrates very clearly how the commitment can be motivationally outweighed simply as a function of the temporal distance of

---

[7] Admittedly, Holton explicitly states that he is describing *full-blown* reconsideration of a resolution, but no indication is given of what would count as less than full-blown reconsideration. It seems from the context that any reconsideration, full blown or otherwise, must involve withdrawing or suspending a commitment.

SS and LL, respectively – and so how one can act against a commitment without reconsidering it.

Even when people in the paradigm case do seem to be reconsidering their commitments, this can often be ex post rationalization of human weakness rather than the ex ante first step of the process of succumbing to temptation. As people fall into the grip of temptation, they can try to justify themselves by finding fault with the commitment they are in the process of abandoning. But that can often be an intellectualized and self-deceiving confabulation about how one got to an outcome rather than part of the process that delivered that outcome. Holton is perfectly aware, of course, of the power of temptation. Here is what he says

> Although to reconsider a resolution is not the same as to revise it, it can be hard to keep the two separate. For when temptation is great its force will quickly turn a reconsideration into a revision. Suspending a resolution can be like removing the bolts on a sluice gate: although one only meant to feel the force of the water, once the bolts are gone there is no way of holding it back. (Holton 2009: 122)

Temptation typically strikes, Holton seems to be saying, once a resolution has been suspended. More often, though, resolutions are suspended because temptation strikes, rather than the other way around. Or at least, I would suggest, that is how things work in the common-or-garden paradigm case. No doubt things were different for St. Anthony the Great in the desert or for St. Ignatius of Loyola (whom Holton quotes in this context), but the phenomenon I am trying to understand is considerably more quotidian.

A second concern with Holton's account is whether it really offers an account of success in the paradigm case. In other words, resolution does not seem to give us a model of resolute choice. The resolute chooser is someone who arrives at the choice node at time t_2 and chooses in accordance with his earlier commitment, thereby overriding the (anticipated but temporary) attractiveness of SS. Of course, therefore, the resolute chooser has to make a choice. Both SS and LL have to be open to him at time t_2. If SS and LL are not both live possibilities at the moment of choice, then it is not really a choice. If SS has somehow been removed from the field of play, then in the last analysis we have a version of the sophisticated choice strategy. The agent has found a way of tying himself to the mast in order to avoid confronting the reality that SS motivationally outweighs LL.

But this seems to be what is going on with an agent who exercises strength of will to avoid rationally reconsidering his resolution. It is hard to see how one would be able to avoid rationally reconsidering one's earlier resolution were one to weigh LL in the balance against SS at the moment of choice. After all, to take SS to be a live option is to take it to be a live option to abandon one's earlier commitment. By Holton's own lights, one cannot take it to be a live option to abandon an earlier commitment without suspending it (removing the sluice bolts!). And so, contraposing, an agent who is avoiding rationally reconsidering his resolution cannot take SS to be a live option, which would seem to make him a sophisticated chooser rather than a resolute chooser.

Let me turn now to the final problem for Holton's account. This emerges when we ask what happens when a resolution is rationally reconsidered. Holton focuses on how rationally reconsidering a resolution can lead to its being revised (as in the sluice gate metaphor). But, of course, as he points out, rational reconsideration is not the same as revision. There must be circumstances when a resolution is reaffirmed, not rejected, as a consequence of being rationally reconsidered. And so, since the process is one of *rational* reconsideration, there must be an account of when such reaffirmation would be rational. This is not an issue that Holton addresses, however. He has interesting things to say about when it would be rational to start the process of reconsidering a resolution (and hence when it would be rational to suspend the resolution), but not about what would make it rational to retain (or reject) a resolution that one was rationally reconsidering.

And yet this is precisely the problem that a satisfying account of resolute choice needs to solve. The resolute chooser finds herself at the choice node at time t_2. In order to make a decision, she needs to weigh up the short-term attractiveness of SS and compare it with the long-term attractiveness of LL. Since she is doing this, she must (as we have already seen) be rationally reconsidering her resolution/commitment to LL. And so to say that it is rational for her to choose LL is in effect to say that she is rationally reaffirming her resolution to choose LL. But then we need to know what makes that reaffirmation rational.

Drawing the threads together, therefore, we can extract three desiderata for an account of self-control from this discussion of Holton. A fully satisfying account

1. Must give a psychologically realistic account of both what it is to be threatened by temptation and what it is to resist that temptation.
2. Must not collapse into a version of sophisticated choice.

3. Must explain how a resolution can be rationally reconsidered and reaffirmed.

The next section returns to the earlier discussion of framing in order to offer an account of self-control that meets these three desiderata. It will not, however, be an account that develops McClennen's ideas about resolute choice. On the contrary, the account I will propose preserves (a form of) the separability principle.

8.4 The Framing Model of Self-Control

As discussed earlier in the context of McClennen, the problem that self-control poses for orthodox decision theory is a function of decision theory's commitment to some version of the separability principle. If, as seems plausible, we equate the exercise of self-control with the capacity for resolute choice, then commitment to separability seems to rule it out as a possibility, leaving myopic and sophisticated choice as the only options. The problem is that in a head-to-head comparison at the moment of choice, separability guarantees that SS will win out over LL because it is given that SS motivationally outweighs LL at time t_2. But to take away the motivational priority of SS at t_2 would be to transform the situation so that it is no longer a paradigm case of self-control.

The problem, really, is that a satisfactory account of self-control that respects something like the separability principle seems to require it simultaneously to be true that SS motivationally outweighs LL and that LL motivationally outweighs SS. The first requirement follows from our description of the paradigm case and the second because separability requires the agent's choice at t_2 to be determined by his preferences at that time. This certainly seems to be a somewhat self-contradictory state to be in, and one might reasonably wonder whether it is even psychologically possible. And even if it is psychologically possible for SS simultaneously to be more and less desirable than LL, how could any such set of preferences be rational? So there are two questions to answer, which I will call the *psychology question* and the *rationality question*.

One obvious way of answering the psychology question would be to invoke some sort of framing effect.[8] If we think about how the options might be

---

[8] I am not aware of appeals to framing in philosophical discussions of self-control, but Al Mele has emphasized the importance of how the objects of desire are represented and evaluated in his discussions of what I am calling the *psychological question of self-control.* See, for example, Mele (1987: 70–72) and Mele (2012: chap. 5), as well as his contribution to this volume (Chapter 9). Our approaches are

framed at the moment of temptation, it becomes much easier to see how LL might be simultaneously preferred to and outweighed by SS. One possible scenario (and, I submit, a very common one) is that the agent prefers LL to SS under one way of framing LL but prefers SS to LL when LL is framed differently. For the discussion that follows, I will assume for convenience that it is LL that is multiply framed. But no loss of generality is intended. There are complementary scenarios where SS is multiply framed – and, of course, where both are.

Let me put some flesh on the bones. On the standard way of looking at things, the agent at time t_2 is faced with two options, a_3 and a_4. The certain outcome of a_3 is LL, and the certain outcome of a_4 is SS. Because SS will be reached sooner than LL, let me make this even more explicit by terming these two outcomes *delayed LL* and *immediate SS*, respectively. Correspondingly, we can label the two options *wait for LL* and *take SS*. At time t_2, the agent strictly prefers *immediate SS* to *delayed LL* and so chooses *take SS* over *wait for LL*. Only the most determined desiderative contortionist could simultaneously (strictly) prefer *delayed LL* to *immediate SS* and (strictly) prefer *immediate SS* to *delayed LL*. However, because each option leads with certainty to its outcome, strictly preferring *immediate SS* to *delayed LL* rules out choosing *wait for LL*.

If we incorporate frame sensitivity into how we think about the moment of choice, though, then we can envision a richer decision problem. Suppose that the agent frames the *delayed LL* outcome in a different way – as *having successfully resisted SS*, for example. Then we have a new strategy available at time t_2, namely *resisting SS*. Let me call it a_5. And this opens the door for preferences that make possible the exercise of self-control while preserving separability. An agent confronting temptation might strictly prefer *having successfully resisted SS* to *immediate SS*, at the same time as strictly preferring *immediate SS* to *delayed LL*. Given these preferences, one would expect the agent to choose the option leading to her most preferred outcome, which would, of course, be a_5 (*resisting SS*). Because *having successfully resisted SS* and *delayed LL* correspond to two different ways of framing the outcome of the agent eventually receiving LL, the agent does indeed successfully

complementary, although mine is more explicitly developed within the context of a broadly decision-theoretic approach to practical reason and rationality. Frederic Schick has emphasized the general importance of frame sensitivity in a series of very insightful books (although he does not use the terminology of framing, preferring instead to talk about understandings, and he does not really engage with self-control). See Schick 1991, 1997, 2003. Bermúdez (forthcoming: chap. 6) discusses some of the parallels and differences between my way of thinking about frames and Schick's views about understandings.

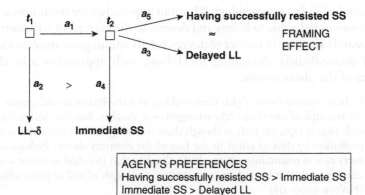

Figure 8.6 The framing model applied to the paradigm case of self-control

implement the plan that she formed at time t_1. And so she exercises self-control.

However, the framing model is not a way of developing McClennen's notion of resolute choice. This is because the framing agent implements her plan in a way that is consistent with the separability principle. The separability principle says that at a particular choice node the agent should be guided only by her preferences over the options and payoffs available at that node. And that is certainly true here, as emerges clearly when Figure 8.6 is compared with Figure 8.5.

Resolute choice depends on the agent's preferences at time t_1 somehow being effective at time t_2 despite the preference reversal that takes place between t_1 and t_2. In contrast, the framing model has no need to go beyond the preferences and options available at the moment of choice.

Because the framing model retains a commitment to separability, it has no need to appeal to pragmatic arguments of the type that McClennen brings into play to secure the rationality of choosing resolutely. If indeed it is rational to exercise self-control in the way the model describes, then rationality is secured by straightforward maximizing considerations, not pragmatic ones that involve taking into account the preferences of past and/or future selves. I turn to the rationality question in the next section. Before that, however, let me return briefly to Holton.

One very interesting aspect of Holton's discussion of self-control is his highlighting of how self-control involves intrapersonal, psychic conflict. It is because they cannot do justice to the conflict dimension that Holton rejects attempts to accommodate self-control within the ambit of a broadly

Humean belief-desire psychology. Humean approaches see intentions and resolutions as reducible to beliefs and desires, so the exercise of self-control is ultimately a matter of one set of desires/beliefs winning out over another set of desires/beliefs. According to Holton, such approaches miss the essence of the phenomenon.

> If these accounts were right, then sticking to a resolution would consist in the triumph of one desire (the stronger) over another. But that isn't what it feels like. It typically feels as though there is a *struggle*. One maintains one's resolution by dint of effort in the face of the contrary desire. Perhaps not every case of maintaining strength of will is like that (we shall mention some that are not). But by and large maintaining strength of will requires effort. (Holton 2009: 118)

On Holton's view, the sense of struggle emerges from the battle to resist rational reconsideration of one's resolutions. I propose an alternative explanation. The characteristic struggle of self-control reflects a clash between different ways of framing a given outcome. The agent is struggling between different ways of framing one or more of the relevant outcomes. This is an exercise that can certainly require effort. It is no easy matter to shift one's perspective away from immediate gratification and toward the virtue of resisting temptation.

Because frames are not themselves reducible to beliefs or desires, this suggestion preserves Holton's anti-Humean insight. At the same time, because how one frames an outcome is driven by emotional and other affective engagements with it (more on this in the next section), the framing model offers a less intellectualized picture of psychic conflict and so is more psychologically realistic.

Let me return briefly to the three desiderata identified in this section for an account of self-control.

1. It must give a psychologically realistic account of both what it is to be threatened by temptation and what it is to resist that temptation.
2. It must not collapse into a version of sophisticated choice.
3. It must explain how a resolution can be rationally reconsidered and reaffirmed.

The discussion in this section shows that the framing model meets the first and second of these. As yet, though, we have not tackled the question of rationality. This will be the task of the next section.

8.5   Framing, Self-Control, and Rationality

Can the framing model of self-control explain how a resolution or plan can be rationally reconsidered and affirmed? It certainly has an account of how the resolution or plan can be reconsidered and reaffirmed in the paradigm case. In brief, reconsidering the plan takes place in the context of reframing one or more of the different outcomes. In the preceding section we considered how *delayed LL* might be reframed as *having successfully resisted SS*, but, of course, it is equally possible for *immediate SS* to be reframed (as *abandoning LL in the face of temptation*, for example). But the key question at hand is whether such a process can be rational (given that it is perfectly clear to the agent that in each case these are different ways of framing the same outcome).

I will follow two separate strategies for arguing that to revert to the terminology of Section 8.2, valuation is an ultraintensional context. First, I will offer three very general sorts of reasons for thinking that frame-sensitive valuations (and, by extension, frame-sensitive reasoning) can, in certain circumstances, be perfectly rational. There is nothing in these general reflections that is specific to self-control. The second strategy, in contrast, argues that the reasons for which an agent might arrive at the kind of preferences characteristic of self-control in the paradigm case are ones that *can* be rational. This will require looking more closely at the mechanisms of framing.

Here are the three general reasons:

1. **Due diligence.** Rationality imposes informational requirements of due diligence. A rational decision maker must look through an action to its consequences. In many cases, though, consequences are not simply given. What consequences an act is judged to have is a function of how that act is framed. And the consequences themselves can be framed in different ways. Thus, exercising due diligence can lead a rational decision maker to think about an act and its consequences under multiple frames.
2. **Perceived similarities.** We often value things as a function of how similar they seem to other things that we value. But similarities are not simply given. The similarities that we see between things are in many ways determined by how we frame them. Consider genetically modified organisms (GMOs), for example. We can consider GMOs under a husbandry frame (as a natural next step from millennia of selective breeding) or under a monopoly capitalism frame (as ways in which unscrupulous corporations can hold agriculturalists hostage). Shifting

frame can shift the similarity classes, and hence the valuations, in ways that a rational decision maker might well want to take into account.

3. **Emotional engagement.** Valuations often emerge from emotional engagement. As with perceived similarities, how one engages emotionally with an outcome can vary according to how it is framed. As a consequence, frame-dependent emotional engagements can vary in ways that force different valuations. One can think about climate change, for example, under the natural disaster frame or under the frame of one's obligations to future generations. The first frame typically elicits a stronger emotional response, but could a rational decision maker not adopt these different valuations simultaneously?

These three observations are closer to intuition pumps than arguments, of course. But they should at least start breaking the hold of the orthodox view that frame sensitivity is inevitably irrational.

Let me focus now on how framing might work in self-control. There is an obvious problem. If the framing model is correct, then the agent exercising self-control has the following preferences:

α. *having successfully resisted SS* is strictly preferred to *immediate SS*.
β. *immediate SS* is strictly preferred to *delayed LL*.

Since the agent is perfectly well aware that *having successfully resisted SS* is ultimately just a different way of framing *delayed LL*, on the face of it she either has intransitive preferences or she strictly prefers an outcome to itself. Neither seems to be the characteristic behavior of a rational agent.

Plainly, though, there is an equivocation here. The strict preference relation is being taken to range simultaneously over two very different types of things. Suppose that we distinguish between intensional and extensional construals of outcomes. Construed extensionally, an outcome is just a state of affairs in the world – a complex of individuals and properties holding at a time (say the agent having at time t_3 the property of *receiving LL*). Construed intensionally, an outcome is a state of affairs conceptualized (or framed) in a certain way. The agent exercising self-control on the framing model has strict preferences over outcomes construed intensionally, but the apparent irrationality only emerges when outcomes are construed extensionally.

This suggests that we might appeal to a standard maneuver for accommodating putatively rational breaches of transitivity (or of any other of the standard axioms of Bayesian decision theory). This maneuver is sometimes

termed "loading up the consequences."[9] In effect, it rejects the idea that
outcomes can be individuated extensionally, so *having successfully resisted
SS* and *delayed LL* come out as different outcomes. This leaves no perspective from which the agent comes out as either strictly preferring an outcome to itself or as having intransitive preferences.

Applying the standard maneuver, though, seems to eliminate the idea
that we are dealing here with a framing effect. If there is no sense in which
having successfully resisted SS and *delayed LL* correspond to a single outcome, then it does not seem legitimate to describe them as two different
frames. So there needs to be some way in which the two outcomes are
extensionally equivalent. And it is not helpful to circumscribe the reach of
transitivity so that it extends only to outcomes as intensionally framed.
This might be a plausible move if we were dealing with a regular intensional or hyperintensional context, but, according to the framing model of
self-control, the agent's preferences at time t_2 reflect a framing effect of
which the agent herself is fully aware. Thus, in addition to the strict
preferences α and β identified earlier, we have to attribute to the agent
knowledge of the identity

γ. *having successfully resisted SS = delayed LL*

It is because we have α, β, and γ together that I have described valuation as
an ultraintensional context. But then the standard maneuver is surely
blocked, and the earlier argument would seem to be back in play.

Let's look at the argument a little more carefully, though. In the background is something like the following thought: if the agent knows that
having successfully resisted SS = delayed LL, then that knowledge imposes
what John Broome has termed a "rational requirement of indifference."[10]
It is a requirement of rationality, in other words, to be indifferent between
two outcomes known to be extensionally equivalent. But why should this
be? The obvious answer is that outcomes known to be extensionally
equivalent cannot differ in ways that could make it rational to have
a preference between them. As Broome puts it,

> [i]f rationality requires one to be indifferent between two alternatives, then
> the alternatives do not differ in a justifier: they do not differ in any way that
> justifies a preference between them. (Broome 1991: 104)

---

[9] See, for example, Risse (2002).
[10] See Broome (1991: 103–6) for a discussion of rational requirements of indifference in the context of
potential counterexamples to transitivity.

This suggests a strategy. To show that the agent exercising self-control is rational in valuing *having successfully resisted SS* differently from *delayed LL*, we need to find rationally relevant differences between them.

That there could well be such rationally relevant differences between them is strongly suggested by some of the considerations that emerged at the beginning of this section. We engage with the world as we frame it, not as we find it. To put it in slightly different terminology, we engage with outcomes under modes of presentation, not as they are "in themselves." Different frames can suggest different comparison classes and, as a consequence, elicit different emotional reactions. All these things can lead to different valuations of outcomes known to be extensionally equivalent. That there will often be such different valuations seems psychologically unavoidable. This does not, of course, make them rational, but it at least allows us to formulate the question at hand more precisely. The question of whether valuation is an ultraintensional context is really this: could such frame-derived differences in valuation be rationally justified?

To make the case for an affirmative answer here, I will propose some ways of thinking about the differences between *having successfully resisted SS* and *delayed LL* that seem (to me at least) to make discriminating between them rationally justifiable. To organize the discussion, I distinguish three different types of reason-giving considerations, borrowing from Robert Nozick's distinction in *The Nature of Rationality* among three different ways of thinking about utility (Nozick 1993). The first set of reason-giving considerations is *causal*, the second *evidential*, and the third *symbolic*. Nozick's own use of this three-way distinction is somewhat different. He wants to discuss three different types of utility. From my perspective, however, these are different ways of thinking about an outcome that are made available by how it is framed.

1. **Frame-sensitive causal reasons.** Framing the relevant outcome as *having successfully resisted SS* rather than as *delayed LL* opens up a range of ways of thinking causally. It is a familiar idea (promoted by George Ainslie and others) that the exercise of self-control can be self-reinforcing. The simple fact that I exercise self-control now strengthens my willpower and so makes it easier for me to exercise self-control in the future. It seems to me, moreover that it is something that can potentially serve as a reason for an agent in the paradigm case.

It is a reason counting in favor of holding out for LL that it will strengthen one's willpower and so lessen the struggle and effort required to secure the long-term goal in comparable situations in the future. And so it should be something that can feature in an agent's practical reasoning, most obviously as reflected in how she evaluated the outcome. The causal power of exercising self-control can add value, one might say, to the outcome of holding out for LL. But in order for this to work, the outcome needs to be framed in the right way. The *having successfully resisted SS* frame seems much more likely to lead the agent to be sensitive to the causal powers of exercising self-control than the *delayed LL* frame.

2. **Frame-sensitive evidential reasons.** One of the lessons of Newcomb-type cases is that we need to distinguish sharply between an action causally contributing to some further outcome and its providing evidence for (increasing the probability of) that further outcome occurring. In the paradigm case, of course, choosing option a_3 (or the equivalent a_5) will lead to the LL outcome with certainty. But it can also provide diagnostic evidence. I might be skeptical, for example, of the view just sketched out in the preceding paragraph. I might not believe that my exercising self-control now can have any effect at all on what happens in battles with temptation. But still, I might nonetheless believe that my exercising self-control on this occasion is good evidence that I will exercise self-control in the future.

There are broader evidential dimensions to exercising self-control. Resisting temptation can provide evidence about the sort of person I am – about my character traits and strength of will. For most of us, positive information here would be welcome news. This potential for bringing welcome news should be reflected in how we value outcomes. Part of the attractiveness of the outcome of attaining LL is its evidential significance. And yet, as with the causal dimension just discussed, the evidential dimension seems much more likely to emerge vividly on the *having successfully resisted SS* frame than on the *wait for LL* frame.

3. **Frame-sensitive symbolic reasons.** Consequentialist approaches to value are often criticized for neglecting the "intrinsic" aspect of actions. This is, however, a mistake. One of the consequences of performing an action is that one has performed an action with such-and-such an intrinsic aspect, which is therefore completely available to a rational decision maker to take into account. This holds in particular for the symbolic dimension of an action. Actions can be valued for what they symbolize. A religious person, for example, might value the

religious symbolism of the act of resisting temptation (as symbolizing the very same battle won so triumphantly by St. Anthony in the wilderness, for example).

This symbolic value is, of course, neither causal nor evidential. One can be moved by an action's symbolic value without thinking that the action will bring about that which it symbolizes or even increase the evidence for it. And, particularly in the case of decision making under certainty, the symbolic value of an action will be reflected in how one values the outcome to which it leads. But, as with the causal and evidential dimensions, such valuation is more likely to emerge on some framings than others.

In sum, there is a range of frame-sensitive factors that are rationally relevant to how an agent values outcomes. One might reasonably expect them to emerge more clearly on some frames than on others. In particular, it seems perfectly reasonable to think that they might rationally attach to an outcome when it is framed one way, but not when it is framed another way – even when the agent is aware that the two frames correspond to a single outcome. If all that is true, then valuation is an ultraintensional context.

Recognizing the possibility of frame-sensitive valuations provides a powerful explanation of how self-control can be exercised in the paradigm case. This explanation is an alternative to resolute choice and satisfies the desiderata that emerged from considering Holton's model of self-control. Moreover, there are good reasons, I claim, for thinking that valuation is an ultraintensional context, which means that these frame-sensitive divergences in valuation can be perfectly rational.

References

Arrow, K. J. 1982. Risk perception in psychology and economics. *Economic Inquiry* 20:1–9.

Bermúdez, J. L. Forthcoming. *The Power of Frames: New Tools for Rational Thought*. Cambridge: Cambridge University Press.

Broome, J. 1991. *Weighing Goods: Equality, Uncertainty, and Time*. Oxford: Basil Blackwell.

Holton, R. 2009. *Willing, Wanting, Waiting*. New York, NY: Oxford University Press.

Kühberger, A. 1995. The framing of decisions: A new look at old problems. *Organizational Behavior and Human Decision Processes* 62:230–40.

McClennen, E. F. 1990. *Rationality and Dynamic Choice*. Cambridge: Cambridge University Press.

Mele, A. R. 1987. *Irrationality: An Essay on Akrasia, Self-Deception, and Self-Control*. Oxford: Oxford University Press.

Mele, A. R. 2012. *Backsliding: Understanding Weakness of Will*. New York, NY: Oxford University Press.

Nozick, R. 1993. *The Nature of Rationality*. Princeton, NJ: Princeton University Press.

Risse, M. 2002. Harsanyi's "utilitarian theorem" and utilitarianism. *Nous* 36: 550–77.

Schick, F. 1991. *Understanding Action*. Cambridge: Cambridge University Press.

Schick, F. 1997. *Making Choices*. Cambridge: Cambridge University Press.

Schick, F. 2003. *Ambiguity and Logic*. Cambridge: Cambridge University Press.

Tversky, A., and D. Kahneman. 1981. The framing of decisions and the psychology of choice. *Science* 211:453–58.

CHAPTER 9

Exercising Self-Control
An Apparent Problem Resolved

Alfred R. Mele*

In a project description that accompanied his invitation to contribute to this volume, José Bermúdez offered a definition of "a paradigm case for discussions of self-control" and asked two questions about the paradigm case. He wrote:

> The paradigm case occurs when an agent makes at time t_1 a commitment or resolution to pursue a large, long-term benefit (henceforth: LL) at a later time t_3. At a time t_2, later than t_1 and earlier than t_3, the agent has the opportunity of a small, short-term reward (henceforth: SS). Although at the time of making the resolution the (discounted) value of LL is more powerfully motivating than the (discounted) value of SS, by t_2 the agent's preferences have temporarily reversed and now SS motivationally outweighs LL.

His questions were these:

1. How is it possible to exercise self-control in the paradigm case?
2. When, how, and why is it rational to exercise self-control in the paradigm case?

My primary aim in this chapter is to answer a version of question 1. I say "a version" because I make some modest modifications to Bermúdez' description of the paradigm case and add an assumption that makes the question more challenging than it would otherwise be. The question is answered in Section 9.1. In the remainder of the chapter I comment briefly on my "judgment-involving" conception of mainstream exercises of self-control and on Bermúdez' second question.

---

* I am grateful to an audience at Texas A&M University for discussion and to José Bermúdez for comments on a draft. Parts of this chapter derive from Mele (2012). This chapter was made possible through the support of a grant from the John Templeton Foundation. The opinions expressed here are my own and do not necessarily reflect the views of the John Templeton Foundation.

9.1   Exercising Self-Control in a Paradigm Case

I modify Bermúdez' description of the paradigm case in four ways. First, I add that the agent judges it best on the whole, from the perspective of his own values and beliefs, to pursue LL and does not change his mind about that.[1] Second, I reformulate talk of the motivational strength of a (discounted) value in terms of the motivational strength of a desire. Third, I make it explicit that achieving SS is incompatible with achieving LL. Fourth, because preference is understood in different ways by different theorists, I avoid using the term *preference*.[2] The result reads as follows:

> The paradigm case occurs when at a time t_1 an agent judges it best on the whole (from the perspective of his own values and beliefs) to pursue a large, long-term benefit (LL) at a later time t_3 and makes at t_1 a commitment or resolution to pursue LL at t_3. At a time t_2, later than t_1 and earlier than t_3, the agent has the opportunity of a small, short-term reward (SS). Although at the time of making the judgment and resolution his desire to pursue LL is motivationally stronger than his desire to pursue SS, and although at t_2 the agent believes that it is better to pursue LL than SS, by t_2 SS motivationally outweighs LL. Getting either of LL or SS precludes getting the other.[3]

To make things more difficult for myself than they would otherwise be, I make the following background assumption:

> *T*: Whenever we act intentionally, we do, or try to do, what we are most strongly motivated to do at the time.[4]

Motivational strength, as I understand it, is a matter of causal power. Other things being equal, if an agent's desire to *A* is stronger than her desire to *B*, the former desire has more power to issue in her *A*-ing than the latter does to issue in her *B*-ing.[5]

---

[1] On such judgments, see Mele (2012: 4–7).

[2] If preferring *x* to *y* is just a matter of *x* being "more powerfully motivating" than *y*, then "the agent's preferences have ... reversed" in Bermúdez' description of the paradigm case is redundant and can safely be removed.

[3] Should someone complain that I have shifted the paradigm case outside the realm of classical decision theory, I would reply that the paradigm case, as I have reformulated it, also merits philosophical attention. For more on this, see Section 9.2.

[4] I defend qualified versions of this thesis in Mele (1992: chap. 3; 2003: chap. 8). An undemanding notion of trying is common in the philosophy of action literature. It is the notion at work here. Although people may often reserve attributions of trying for instances in which an agent makes a considerable or special effort, this is a matter of conversational implicature and does not mark a conceptual truth about trying. As I understand trying, when, just now, I typed the word *now*, I tried to do that even though I easily typed the word.

[5] For articulation and defense of a notion of motivational strength, see Mele (2003: chaps. 7 and 8). The motivational strength of a desire should not be confused with its affective intensity. For example, "Carol, a morally upright psychiatrist who experiences, to her own consternation,

By t_2, "SS motivationally outweighs LL." A theorist may contend that as a consequence, by this time it is too late for the agent to bring it about that she pursues LL at t_3. Such a theorist might offer the following argument for this contention:

1. Because SS motivationally outweighs LL at t_2, the agent's motivation at t_2 to exercise self-control in the service of LL is outweighed by her motivation not to do that.
2. Whenever we act intentionally, we do, or try to do, what we are most strongly motivated to do at the time. (Assumption T.)
3. So the agent will not intentionally exercise self-control in the service of LL at t_2.
4. Nor will the agent nonintentionally exercise self-control in the service of LL at t_2.
5. So the agent will not bring it about that she pursues LL at t_3.

Assume that premise 2 – that is, T – is true. Should we be persuaded by this argument?

Having a particular case before us will prove useful. Consider the following one[6]: Ian has been doing some outdoor painting at his daughter's house on a hot summer day. He decided to take a break for lunch and to watch TV while eating. When he finishes his lunch, Ian thinks about how much longer he should put off getting back to work. Although he does not often watch golf on TV and generally regards doing so as boring, this is what he has been doing, and he finds it much more enjoyable than painting in the hot summer sun. After some thought, Ian judges it best to shut off the TV and get back to work when the next commercial break starts. Finishing the painting job by 5:00 is important to him because he promised his daughter he would do that, and he is convinced that he needs to get back to work very soon if he is to finish on time. Ian comes to his conclusion during a commercial break, and he resolves to act accordingly. About 10 minutes later, Ian can tell that ads are about to start running. By this time (t_2), he is more strongly motivated to continue watching TV through the next ads and beyond than he is to get back to work when the ads start.

According to proponents of the argument at issue, Ian will not at t_2 exercise self-control to bring it about that he gets back to work when the

a phenomenologically intense urge for a sexual romp with a seductive patient, may have, in a desire that has little phenomenological kick, a stronger inclination to forego that course of action" (Mele 2003: 162).

[6] For a similar story about Ian, see Mele (1987: 69).

ads start. An important plank in their argument is the claim that because continued relaxation motivationally outweighs the longer-term gain at t_2, Ian's motivation at t_2 to exercise self-control at that time to get himself back to work on schedule is outweighed by his motivation not to engage in such an exercise of self-control. Should this plank be accepted? Is it an instance of a more general truth?

Attention to a case with a significantly greater time scale provides a useful perspective on these questions. Here is such a case from Mele (1996: 55–56):

> Wilma, who suffers from agoraphobia, has been invited to her son's wedding in a church several weeks hence. Her long-standing fear of leaving her home is so strong that were the wedding to be held today, she would remain indoors and forego attending. Wilma is rightly convinced that unless she attenuates her fear, she will not attend the wedding. And there is a clear sense in which she is now more strongly motivated to remain at home on the wedding day and miss the wedding than to attend it: her current motivational condition is such that, unless it changes in a certain direction, she will [stay home]. (A nice conceptual test of what one is now most motivated to do at a later time is, roughly, what one would do at that later time if one were then in the motivational condition that one is in now.) Further, Wilma believes that, owing to her motivation to remain in her house indefinitely, she probably will miss the wedding [but she judges that attending the wedding would be much better than staying home].

As I have argued elsewhere (Mele 1996: 57–61), we lack good grounds for maintaining that because Wilma is now more strongly motivated (in the sense identified) to stay home on the wedding day than to attend the wedding, she also is more strongly motivated not to try to attenuate her fear and the associated motivation to stay home than to try to do this. She sees her fear as an obstacle to something she values doing, and the motivational strength of any desire she may have not to try to attenuate it may be far outstripped by the motivational strength of her fear itself and outstripped, as well, by the strength of her desire to try to bring it about that she is in a position to leave her house on the wedding day and attend the wedding.

Now, although Wilma has a few weeks to bring her motivational condition into line with the course of action she judges best, Ian may have less than a minute. But just as Wilma may desire more strongly to try to change her motivational condition than to allow it to persist, so may Ian. And just as Wilma may desire more strongly to try to attenuate her fear in the service of attending the wedding than not to try this, even though her fear currently is motivationally stronger than her desire to attend, Ian may desire more strongly to try to boost his motivation to get back to work on schedule than not to try this, even though his desire to continue relaxing in front of the TV currently is stronger than his desire to get back to work when the ads start.[7]

Readers who find my claims about Wilma plausible may be skeptical of my analogous claims about Ian. The next item of business is to eliminate the skepticism.

Why – that is, for what reason – would Ian exercise self-control to get himself back to work on schedule? Here is a straightforward answer: because he continues to believe that getting back to work on schedule is better, all things considered, than lingering in front of the TV.

How would Ian exercise self-control? Here is one possibility. Imagine that Ian has learned that in similar circumstances, when he is tempted to procrastinate, speaking sternly to himself sometimes works. To say now that these circumstances include his being more strongly motivated to procrastinate than not to do so would be question begging. So I am silent on that particular point. Suppose that, in light of what he has learned, it occurs to Ian to utter a self-command – perhaps this: "Get off your butt, Ian, and get back to work." Then we have a candidate for an answer to my *how* question: namely by uttering a self-command of the kind described.

Is it *possible* for Ian, given his circumstances, to utter this self-command in the service of getting himself back to work on schedule? It might be thought that premise 2 (assumption *T*) yields a negative answer. Someone who takes this view of things may argue for it as follows:

> *T*: Whenever we act intentionally, we do, or try to do, what we are most strongly motivated to do at the time.
>
> *A*: At t_2, Ian is more strongly motivated to continue watching TV then than he is to do anything else then.
>
> *B*: So at t_2 Ian will continue watching TV, and he will not utter the self-command then.

This argument is invalid. *T* neither asserts nor entails that whenever we act intentionally, we perform only one intentional action. *T* leaves it open

---

[7] Both Ian and Wilma have desires about their own motivational attitudes. It can be said that Ian desires that his desire to get back to work before the ads start be stronger than his opposing desire and that Wilma desires that her desire to attend the wedding be stronger than her desire to stay in her house. It should not be inferred from this that I believe that exercising self-control depends on having relevant higher-order desires. For an argument that the dependence claim is false, see Mele (1995: chap. 4). That chapter includes a discussion of Richard Jeffrey's intriguing article, "Preferences among Preferences" (Jeffrey 1974).

that Ian intentionally utters the self-command *while* he is doing what he is most strongly motivated to do at the time – that is, while he is watching TV. Why would Ian utter the self-command given his motivational condition? I have already answered that question. But I have not yet fully answered the question about possibility that I raised.

Even if it is possible for Ian to utter the self-command in his circumstances, is it possible for his attempt at self-control to succeed – that is, to issue in his getting up from his easy chair before the ads begin and getting back to work? Consider the following from B. F. Skinner (1953: 236): "In getting out of a bed on a cold morning, the simple repetition of the command 'Get up' may, surprisingly, lead to action. The verbal response is easier than getting up and easily takes precedence over it, but the reinforcing contingencies established by the verbal community may prevail." In *Autonomous Agents* (Mele 1995: 44–45), I develop this idea at some length in connection with a story like Ian's (also see Mele 1987: 69–72). Here I keep things simple.

Why is it easier, as Skinner says, to command oneself to get out of bed than actually to get out of bed? Well, for one thing, one can utter the command without leaving the comfort of one's bed, but one cannot get out of bed without leaving the comfort of one's bed.

How might the self-command strategy work (if it does)? Here I offer just one hypothesis about that. I need to do some stage setting first.

According to a familiar, traditional way of conceptualizing self-control, it is the contrary of weakness of will, or *akrasia*. I have defined core *weak-willed action* – a pure case of the phenomenon – as "free (and therefore uncompelled), sane, intentional action that, as the nondepressed agent consciously recognizes at the time of action, is contrary to his better judgment, a judgment based on practical reasoning" (Mele 2012: 33). Some philosophers have argued that such actions are impossible. My own view is that actions of this kind are not only possible but also actual – indeed, a familiar part of the real world (Mele 1987, 2012).

My view rests partly on the following two propositions and on various arguments for them.

> P1: Our better judgments normally are based at least partly on our evaluations of objects of our desires (that is, desired items).
>
> P2: The motivational strength of our desires does not always match our evaluations of the objects of our desires.

If both these propositions are true, it may sometimes happen that although we judge it best to A and better to A than to B, we are more strongly motivated to B than to A. And given how our motivation stacks up on these occasions, B-ing – rather than A-ing – is to be expected.

Our evaluations of the objects of our desires are evaluative beliefs about the desired items. For example, Ian may believe that the value of keeping his promise to his daughter far exceeds that of relaxing in front of the TV, and Wilma may believe that the value of attending the wedding is much greater than that of staying home.

P_I is a plank in a standard conception of practical reasoning. In general, when we reason about what to do, we try to figure out what it would be best, better, or good enough to do, not what we are most strongly motivated to do. When we engage in such reasoning while having relevant conflicting desires, our concluding judgments typically are based partly on our evaluations of the objects of those desires – which may be out of line with the motivational strength of those desires if P_2 is true.

P_2 is confirmed by common experience and thought experiments (see Mele 1987: 37–39), and it has a foundation in scientific studies, as I have explained elsewhere (Mele 2012: chap. 4). Influences on desire strength are not limited to evaluations of the objects of desires, and other factors that influence desire strength may fail to have a matching effect on assessments of desired objects.

Regarding P_2, return to my story about Wilma. If the strengths of her pertinent desires (including her desire to go to the wedding and her desire to stay home) had matched her evaluations of the objects of those desires, she would have been much more strongly motivated to attend the wedding, as she judged best, than to stay in her house. There would have been no need for an exercise of self-control. Similarly, as I observe in my book *Irrationality* (Mele 1987: 37), someone with a severe fear of flying may judge it best to board a plane now (because it is the only way to get to an important job interview) and yet be so anxious about flying that he does not board the plane. If the strengths of his desires had matched his evaluations of the objects of those desires, boarding the plane would not have been a problem.

Despite what I have said thus far, some readers may be attracted to the idea that it is a necessary truth that the motivational strength of a desire always matches the agent's evaluation of the desire's object. I invite them to consider the following far-fetched scenario from *Irrationality* (Mele 1987: 37–38):

Imagine that an evil genius is able to implant and directly maintain very strong desires in people, and that he does this to Susan. However, because she knows both that her desire to *A* was produced by the evil genius and that he does this sort of thing solely with a view to getting those in whom the desires are implanted to destroy themselves, Susan gives her *A*-ing a very low evaluative ranking. She believes that her not *A*-ing would be much better, all things considered. Nevertheless, the genius's control over the strength of Susan's desire to do *A* is such that the balance of her motivation falls on the side of her *A*-ing [that is, it is such that she is more strongly motivated to *A* than not to *A*].

This scenario certainly seems conceptually possible.

If we were *ideal* agents, our evaluations of the objects of our desires might always determine and be matched by the strength of those desires. If we were agents like this and we ranked quitting smoking much higher than smoking our next cigarette, studying now much higher than playing video games now, forgoing an after-dinner snack much higher than eating one, and so on, our desires for the more highly ranked conduct would be much stronger than our competing desires, and acting as we judged best would be easy. But there is a lot of evidence that we are not ideal agents of this kind. If we were, there would be no market for self-help books that focus on strategies for resisting temptation. Such resistance would be a piece of cake: we would always be most strongly motivated to do what we judge best.

George Ainslie makes a strong case for the claim that the motivational strength of desires tends to increase hyperbolically as the time for their satisfaction approaches (Ainslie 1992, 2001). If there is not a matching tendency in our evaluations of the objects of our desires, it may often happen that the strength of a desire is seriously out of line with the agent's assessment of its object. When a dieter is looking over his dinner menu, he may give eating dessert later a low evaluation, judge it best not to eat dessert, and have a desire of moderate strength for an after-dinner dessert. But when the dessert menu is delivered to his table after he has finished his low-calorie meal, the strength of his desire for dessert may spike dramatically, as predicted by Ainslie's model. If this happens without his assessment of the goodness or value of eating dessert also spiking dramatically, we have motivation-evaluation misalignment, and he may judge it best not to order dessert while being more strongly motivated to order it than not to order it. (In my book *Backsliding* [Mele 2012: 82–86], I explain how this can happen.)

With this background in place, I return to the last question I asked about Ian. How might his self-command strategy work (if it does)? A closely related question is this: how might his self-command help to bring it about that the balance of his motivation shifts in favor of his shutting off the TV when the ads begin and getting back to work?

In *Backsliding* (Mele 2012: 77–82), I review evidence that how one represents the objects of one's desires has an important effect on the strengths of one's desires. We have known this for a long time. When children are shown slide-presented images of reward objects (Mischel and Moore 1973), they hold out much longer for their preferred rewards than they do when they see images of unavailable treats, blank slides, no slides, or (in an earlier study; Mischel and Ebbesen 1970) the rewards themselves. Why is this? Walter Mischel and Ozlem Ayduk (2004: 114) write:

> It became clear that delay of gratification depends not on whether or not attention is focused on the objects of desire, but rather on just how they are mentally represented. A focus on their hot features may momentarily increase motivation, but unless it is rapidly cooled by a focus on their cool informative features (e.g., as reminders of what will be obtained later if the contingency is fulfilled) it is likely to become excessively arousing and trigger the "go" response.

The general idea applies to Ian's self-command. That command may call attention to what is to be gained by getting back to work now and pull attention away from the unpleasantness of the chore and the pleasantness of relaxation. This may boost Ian's motivation to get back to work while dampening his motivation to say seated in front of the TV. Also, if Ian is disposed to obey his self-commands about immediate actions (as suggested by the passage from Skinner [1953] that I quoted), that disposition may contribute to his motivation to act accordingly.

Obviously, I am not suggesting that uttering self-commands is a foolproof way to get oneself to do what one believes one should do. Only a fool would suggest that. What led me to this point was a pair of related questions. How might Ian's self-command strategy work (if it does)? And how might his self-command help to bring it about that the balance of his motivation shifts in favor of his shutting off the TV when the ads begin and getting back to work? I have offered an answer that finds support in work by Mischel and others.

My primary question in this section has been a version of a question raised by Bermúdez. My description of the paradigm case is a modified version of his, as I explained, and I brought in assumption T to make things more difficult for myself. In answering the question, I used a self-command strategy as my primary illustration because it is a simple strategy, and I wanted to keep things simple. Readers should not infer that I believe that this is the only or best self-control strategy. In some cases, when one needs to act fast to bring the balance of one's motivation into line with one's better judgment, it may be the best strategy. In others, it definitely is not. Wilma, for example, three weeks before the wedding might well benefit from exposure therapy, and she might be sufficiently motivated to give it a try in the service of her better judgment.[8] Simply commanding herself, then, to go to the wedding three weeks later does not promise to be effective.

Readers also should not infer that I believe that T is true. One source of skepticism about T is the thought that some intentional actions may be indeterministically caused by their proximal causes – a thought that receives considerable attention in the literature on event-causal libertarian views of free will. If this is so, an agent's being most strongly motivated to A straightaway may make it highly probable that she will straightaway try to A if she does anything intentionally straightaway while permitting some chance that the she will B intentionally and not try to A.[9] This is something that I leave open, and I have articulated variants of T that accommodate the idea (Mele 1992: chap. 3; 2003: chap. 8).

Some readers may worry that T somehow fails to do justice to rational agency. Consider the following strengthened version of T:

> T^*: Whenever we act intentionally, we do, or try to do, what we are most strongly motivated to do at the time, and we neither do nor try to do anything that is supported by contrary motivation.[10]

It may be thought that, other things being equal, an agent whose behavior at a time is consistent with T^* is more rational than an agent whose behavior at a time is consistent with T but not with T^*. Should this thought be endorsed?

Return to Susan's case and imagine that she knows that if she asks a certain enemy of the evil genius for help, there is a chance that this powerful being will force his evil enemy to undo what he did to her so that her strongest motivation is no longer to A. No strings are attached to such

---

[8] Interesting work on exposure therapy includes Back et al. (2001) and Emmelkamp et al. (2002). I discuss exposure therapy in Mele (2014).

[9] Even so, T is more promising than some other propositions that have been used to connect motivation to action in puzzles about self-control. On this, see Mele (2014: 360–62).

[10] José Bermúdez suggested that I comment on something like T^*.

a request, as Susan knows. Her asking for help is supported by motivation that is contrary to Susan's strongest motivation at the time, and even so, asking for help is, by Susan's own lights, far superior to simply allowing the evil genius to have his way. Seemingly, the rational course of action for Susan to take is the one she rationally regards as far superior to the alternatives.

What about Ian? Given his beliefs and desires, his uttering the self-command would make good sense, even though his doing so is supported by motivation that is contrary to his strongest motivation at the time. Seemingly, he would rationally regard his not making any effort at all to resist temptation as an option that is inferior to his attempting to get himself to do what he believes it would be best, all things considered, to do. The same goes for the man in the example by Skinner that I mentioned earlier, if it is assumed that he judges it best to get out of bed then.

If we were ideal agents whose evaluations of the objects of our desires always determined and were matched by the strength of those desires, T^* might be true of us.[11] Moreover, in that case, T^* might do more justice to our rational agency than T does. But we are not such agents, or so I have suggested here and argued at some length elsewhere (Mele 1987, 2012).

9.2 Paradigms and Rationality

What one regards as "a paradigm case for discussions of self-control" may be influenced by such things as what drew one to the topic of self-control in the first place and what sort of work on self-control one is most interested in. I was drawn to the topic (as an undergraduate) by Plato's and Aristotle's work on weakness of will (or *akrasia*), which they regarded as the contrary of self-control (*enkrateia*). Their conception of weakness of will, on one interpretation, features evaluative judgments, as does their conception of self-control.[12] In this section, I comment on my own judgment-involving conception of paradigmatic cases of self-control and on Bermúdez' second question.

As I observe in my book *Autonomous Agents* (Mele 1995: 71, italics altered), traditional conceptions of akratic action and its contrary "revolve around a certain species of commitment to action – the sort of commitment constituted by a ... better ... judgment. An agent who ... judges it best to *A* is thereby *rationally committed* to *A*-ing, in the sense that (as long

---

[11] Why "might"? Recall my remarks on indeterminism a few paragraphs ago.
[12] See, for example, Aristotle (1915: book 7, chaps. 2 and 3).

as the judgment is retained) the uncompelled, intentional performance of any action that he believes to be incompatible with his *A*-ing would open him to the charge of irrationality." An uncompelled intentional action of this kind, I claim, "would be at least *subjectively* irrational – irrational from the agent's own point of view. For, while explicitly holding the judgment, an agent cannot rationally take himself to have, from his own point of view, better (or equally good, 'best' being understood exclusively) grounds for not *A*-ing."

My reformulation of Bermúdez' definition of "the paradigm case" of self-control places the paradigm case in the sphere of self-control, as traditionally conceived. But it occupies only a portion of that sphere. It is not a requirement on mainstream exercises of self-control that the agent at some time faces a temptation that "motivationally outweighs" the motivation he has for *A*-ing, where *A*-ing is what he judges it best to do.[13] Why would an agent exercise self-control in the service of his *A*-favoring better judgment even though he is at no time dealing with such a temptation? Perhaps because he anticipates that unless he takes preventive measures, his balance of motivation will shift in favor of the tempting course of action and make it harder for him to act as he judges best. (Depending on how well Ian knows himself, we can imagine him deciding to eat lunch at the kitchen table without turning on the TV so as to reduce the risk that he will become sorely tempted to procrastinate.)

When we add the motivational feature that is missing in such a case – a temptation that is motivationally stronger than the agent's motivation for the course of action he judges best – we get an interesting puzzle about self-control. In this chapter, drawing on earlier work of mine (Mele 1987, 1995, 2012), I have presented a solution to that puzzle.

I close with a comment on Bermúdez' second question: when, how, and why is it rational to exercise self-control in the paradigm case? I am on record as claiming that "in paradigmatic cases of akratic action, the judgment that the agent acts against is rationally made and rationally held" (Mele 2012: 10). A presumption of rationality in these connections is typically in place in discussions of mainstream akratic action. This is not to say that the discussants all understand rationality in the same way. Nor is it to say that I myself have defended a particular conception of rationality; I have not. In any case, if we base our conception of paradigm cases of self-controlled actions on paradigm cases of akratic actions, we may add to my

---

[13] Mainstream exercises of self-control are made in the service of the agent's better judgment. On another kind of exercise of self-control – an "unorthodox" kind – see Mele (2012: 15–17, 94–96).

description of the paradigm case of self-control that the initial judgment mentioned there is *rationally* made and that at t_2 the agent *rationally* believes that it is better to pursue LL than SS.

An agent may have a rational belief with the following content: it would be better to pursue LL than SS if it were not for the cost to be paid for employing any promising self-control strategy. For example, the agent may believe that to succeed in holding out for LL, she would need to do something on the order of paying someone a large sum of money to monitor her behavior and break her legs should she obtain SS and that once this monetary cost is factored in, it would be better to pursue SS than LL. But suppose that an agent rationally and explicitly believes that it is better to pursue LL than SS, *all things considered* – including the actual or potential cost of making an attempt at self-control. In this case, pursuing SS rather than employing a self-control strategy in pursuit of LL would be subjectively irrational in the sense identified earlier: while the agent explicitly believes this, she cannot rationally take herself to have, from her own point of view, better (or equally good) grounds for refraining from employing a self-control strategy in pursuit of LL and instead pursuing SS.

In cases such as Ian's, of course, the cost simply of employing a self-control strategy – that is, independently of its succeeding – is very small. If his strategy fails, Ian will have wasted his breath for a second or so (assuming that he tries just once). If his strategy succeeds, he will secure what he himself rationally regards as the superior reward. All things considered, it would be rational for Ian to make the effort.

Does the notion of rationality at work in my discussion of Bermúdez' second question have a place in a fruitful investigation of self-control? This is a fair question. But I have done enough work for now. It is time to kick back and relax – perhaps while watching TV.

References

Ainslie, G. 1992. *Picoeconomics*. Cambridge: Cambridge University Press.

Ainslie, G. 2001. *Breakdown of Will*. Cambridge: Cambridge University Press.

Aristotle. 1915. *Nicomachean Ethics*. In *The Works of Aristotle*, ed. W. Ross, vol. 9. London: Oxford University Press.

Back, S., B. Dansky, K. Carroll, E. Foa, and K. Brady. 2001. Exposure therapy in the treatment of PTSD among cocaine-dependent individuals: description of procedures. *Journal of Substance Abuse Treatment* 21:35–45.

Emmelkamp, P., M. Krijn, A. Hulsbosch, et al. 2002. Virtual reality treatment versus exposure in vivo: a comparative evaluation in acrophobia. *Behaviour Research and Therapy* 40:509–16.

Jeffrey, R. 1974. Preference among preferences. *Journal of Philosophy* 71:377–91.

Mele, A. 1987. *Irrationality*. New York, NY: Oxford University Press.

Mele, A. 1992. *Springs of Action*. New York, NY: Oxford University Press.

Mele, A. 1995. *Autonomous Agents*. New York, NY: Oxford University Press.

Mele, A. 1996. Motivation and intention. *Journal of Philosophical Research* 21:51–67.

Mele, A. 2003. *Motivation and Agency*. New York, NY: Oxford University Press.

Mele, A. 2012. *Backsliding*. New York, NY: Oxford University Press.

Mele, A. 2014. Self-control, motivational strength, and exposure therapy. *Philosophical Studies* 170:359–75.

Mischel, W., and O. Ayduk. 2004. Willpower in a cognitive-affective processing system. In *Handbook of Self-Regulation: Research, Theory, and Applications*, ed. K. Vohs and R. F. Baumeister (pp. 99–129). New York, NY: Guilford Press.

Mischel, W., and E. Ebbesen. 1970. Attention in delay of gratification. *Journal of Personality and Social Psychology* 16:329–37.

Mischel, W., and B. Moore. 1973. Effects of attention to symbolically-presented rewards on self-control. *Journal of Personality and Social Psychology* 28:172–79.

Skinner, B. F. 1953. *Science and Human Behavior*. New York, NY: Macmillan.

CHAPTER 10

Putting Willpower into Decision Theory
The Person As a Team Over Time and Intrapersonal Team Reasoning

*Natalie Gold**

10.1 Introduction

In decision theory, a standard way of modeling a person who has to make a series of choices over time is as a series of transient agents, who each make choices at particular times. In this framework, a person has a problem of self-control when the course of action that is preferred by an earlier transient agent relies on a choice by a later transient agent that will not be preferred from that later transient agent's point of view. Faced with such a problem, the earlier transient agent can try to influence the later transient agent's behavior by taking action to alter her later self's incentives or limit her later self's opportunities. However, what she cannot do is use willpower. A person who believes that she can simply make a plan and act on it when the time comes is considered "naïve," and her naivety is a cause of bad outcomes. This view is in stark contrast to much research in psychology and philosophy, which assumes that people have willpower and that exercising willpower can be rational and lead to good outcomes.

The model also has a lacuna when it comes to agency. Problems of self-control are caused by conflicts between transient agents; there is no sense of an agent who has interests that extend over time. The transient agents may care about the outcomes of the other transient agents in the series, which may figure in their preference functions. Nevertheless each transient agent

* Work on this chapter was supported by funding from the European Research Council under the European Union's Seventh Framework Programme (FP/2007–2013)/ERC Grant Agreement No. 283849. I thank the project team, Jurgis Karpus and James Thom, for helpful discussions. I also thank the participants at the Texas A&M University Workshop on Rationality and Self-Control for their feedback on this chapter, especially José Luis Bermúdez, who provided written comments, and the participants at the Stanford University Workshop on Varieties of Agency, especially Michael Bratman and Grace Paterson, whose feedback on a companion paper also had the side effect of improving this chapter.

acts on its own transient interests. Related to this, the standard model lacks the concept of an intention that guides behavior over time. To the extent that the model includes intentions, they are merely predictions of future behavior that turn out to be correct. This connects back to self-control because some philosophers have argued that willpower consists in not revising one's intentions (Holton 1999).

I will show how willpower can be introduced into decision theory – and the gap between psychology, philosophy, and economics bridged – by allowing that there can be multiple levels of agency and applying the theory of team reasoning. Team reasoning is an extension of game theory that allows that there can be multiple levels of agency (Sugden 1993; Bacharach 2006). In standard game theory, the only recognized level of agency is the individual, analogous to way that models of the person over time only allow one level of agency, the transient agent. Team reasoning was developed to understand the behavior of individuals in groups. The idea is that not just individuals but also groups can be agents, so rather than asking, "What should *I* do?," the individuals in the group can ask, "What should *we* do?," which can solve problems of coordination and cooperation. Team reasoning extends game theory to allow more than one level of agency. Although the levels of agency in the original applications are the individual and the group, the conceptual apparatus can be applied to other problems. Thus we can use the theory of team reasoning to introduce a second level of agency, the person over time, into the model of intertemporal choice, as well as the transient agents.

Intrapersonal team reasoning can explain why it is rational to exert self-control without abandoning the decision-theoretic framework, and it can provide a basis for incorporating intentions into game theory. After introducing the intertemporal problem, I will highlight some parallels with interpersonal problems of coordination and cooperation in order to motivate the use of team reasoning. Then I will introduce team reasoning in the interpersonal context before showing how it can be applied to the intrapersonal case.

10.2 Decision-Theoretic Models of the Problem of Self-Control

In decision theory, problems of intertemporal choice are often analyzed as if, at each time t at which the person has to make a decision, that decision is made by a distinct transient agent or *time slice*, the person at time t. Each time slice is treated as an independent rational decision maker, so "the individual over time is an infinity of individuals" (Strotz 1955–56: 179).

This does not imply any metaphysical commitments, in particular it is not an endorsement of perdurantism, the view that things really do consist of temporal parts. Rather, it is a natural way of modeling people because the self at a particular time is the locus of choices, experiences, and perceptions.

Choices are events that are located in time. Choices result in experiences, and an experience is also a type of event had by a person at a time or over a temporal interval. An experience can be preexperienced or relived through anticipation and memory; anticipation and memory are also types of experiences, which occur at a time, even though they are one step removed from the initial stimulus.

First-person perception of events is also had by the self at a time. There is some evidence from psychology that when we contemplate our experiences, as we get further away from the present, we tend to take a third-person perspective. When people are asked to imagine a scene from their past or their future, they are more likely to see themselves in the picture rather than seeing the scene as though through their own eyes (Pronin and Ross 2006). Nor are we good at predicting our future preferences and attitudes or what our future experiences will feel like (Lowenthal and Loewenstein 2001; Loewenstein, O'Donaghue, and Rabin 2003; Van Boven and Loewenstein 2005). There is a gap between our knowledge of our current experience and our future.

The time slice model captures the idea that choices are made at a time, by selves that have a first-person perspective on that time and a third-person perspective on the past and future. We can think of the successive selves as involved in strategic interaction or the self as a community of interests (Schelling 1984; Ainslie 1992), which gives a nice framework in which to set up problems of self-control in the face of temptations. Here is an example from a classic paper by O'Donaghue and Rabin (1999): Suppose that you usually go to the movies on Saturday nights. The schedule for the next four weeks is as follows: week 1 is a mediocre movie, week 2 a good movie, week 3 a great movie, and week 4 a Johnny Depp movie, which is best of all. You also have a report to write for work, due within a month, and in order to write it you know that you will have to stay in one Saturday night and must therefore skip a movie. The question is, when do you complete the report?

It seems obvious that the best overall plan is to do the report on the first Saturday. This is the option that would be chosen by a planner who was working out your schedule in week 0. But all that we need to add is a little bit of present bias, with the current time slice favoring itself, for you to miss the Johnny Depp movie. In order to see this, let us suppose, following

O'Donaghue and Rabin (1999), that the valuation of the mediocre, good, great, and Depp movies are 3, 5, 8, and 13 and that the cost of writing the report is just the cost of not seeing the movie that evening. It is plausible that each time slice gives more weight to its own experiences. Imagine that each time slice places double the weight on its own experiences than those of other time slices. In this case, a naive agent, who believes that she can make a plan and act on it when the time comes – even in the face of temptation – will end up missing the Johnny Depp movie. Come the first Saturday night, the time slice, call it t_1, based on her current valuation, judges that she should go to the mediocre movie (which, with double weight, is valued at 6) and skip the good movie next week (valued at 5). But, next week, t_2 finds the agent in exactly the same situation. She justifies to herself why a night out is particularly valuable to her right now, so she chooses to go to the good movie tonight (now valued at 10), believing she has the willpower to skip the great movie next week (currently valued at 8). The same happens with t_3, leading to the situation where the agent is forced to miss the Johnny Depp movie in week 4.

There are three things to note about this example. First, the time slice that writes the report bears a cost (missing the movie that week) for which others get the benefit (they get to go to the movie other weeks). In economic language, there is an *externality* because the agent's choice has consequences that affect other agents. Second, each time slice magnifies the sacrifice that would be made by the agent but does not realize that other time slices will also have a *present bias*. However, naivety about future selves is not essential to the problem. Even a "sophisticated" agent, who has correct expectations about her future present bias and backward inducts accordingly, will procrastinate for a week in Rabin and O'Donaghue's model.[1] Third, the time slices end up with an outcome that is ranked very low by all time slices. Missing the Depp movie is everyone but t_3's worst outcome, and it is t_3's second-worst outcome after writing the report herself and missing the great movie. Conversely, apart from t_1, who prefers that the

---

[1] Reasoning by backward induction: there will be no choice in week 4. If she has not written the report, then she will have to skip the movie. She can also predict that if she gets to week 3 and has not written the report, then she will end up missing the Depp movie. Given that, if she gets to week 2 without having written the report, then her effective choice would be between missing the good movie and missing the Depp movie. According to the model, this is a big enough difference in payoff that t_2 would prefer to skip the good movie in week 2 than to have t_4 miss the Depp movie. So, if she has not written the report by week 2, she will write it in that week. Thus, in week 1, the choice is between skipping the mediocre movie in week 1 and skipping the good movie in week 2, and t_1 prefers to skip the good movie, so she does not write the report in week 1. Come week 2, t_2 writes the report.

report is written in week 2, all the time slices prefer writing the report in week 1.

Decision theory offers two types of resources for an agent who, in week 0, wants to ensure that she will overcome temptation in week 1. The t_0 self can change the incentives faced by her future selves or she can alter her opportunities (Thaler and Sheffrin 1981; Greer and Levine 2006). For example, she could enter into a side bet with a colleague, which entails paying out a large sum if the report is not done in the first week, or she could rip up her week 1 cinema ticket.

These are intended as examples of changing incentives versus destroying options, but there is a very fine line between the two strategies. Many precommitments, which destroy options, may actually be a way of making an option more costly: when one destroys an option, there is usually the possibility of replacing it, albeit at some cost. In the preceding example, the cinema ticket could presumably be replaced; even if the movie is sold out, there would be some price at which another ticket holder would trade.

In decision theory, a person who has a problem of self-control in the face of temptation faces a foreseen preference change. In the model, if we observe someone who does not give in to temptation, then either an earlier time slice took action to change the incentives or else she never had conflicting preferences in the first place.

10.3 The Strange Lack of Self Over Time

Many philosophers and psychologists believe that we have another resource for resisting temptation: willpower. A popular idea is that willpower involves making *resolutions*, or plans whose purpose is to help us overcome temptation (Holton 1999, 2009). Decision theory has some room for plans – if they are incentive compatible, i.e., if all the stipulated choices will maximize utility at the time of choice and therefore a decision maker can predict in advance that he will make them. This reduces planning to prediction. I will show how decision theory can make room for willpower and for plans that guide action. Although my solution includes an explanation of intentions and resolutions, the basic capacity for self-control is intrapersonal team reasoning. Therefore, my solution explains willpower, but in a different manner from Holton's explanation. In fact, the dependency is in the other direction: we cannot make sense of resolutions within decision theory without adding something like my proposed mechanism of willpower.

I motivate my approach by noting an oddity of the standard decision-theoretic model of the agent over time, namely the strange lack of the self over time. The transient agents in the O'Donaghue and Rabin (1999) model put weight on the outcomes of the other time slices. Nevertheless, the preferences in the model are those of the transient time slices; there is no sense in which they can hold preferences qua continuing self over time. Yet we usually think that people have interests that extend over time.

This way of modeling people, without extended interests or extended selves, has some counterintuitive implications. We can see these very clearly using a simple example where the transient agents share a common goal. Take someone who wants to cross the road, from east to west on a two-lane city street. In order to cross the street, he must perform two actions in sequence. In period 1, he must walk from the east side of the street to the middle. Then, in period 2, he must walk from the middle to the west side. From the perspective of conventional decision theory, there are two transient agents, the agent in period 1 (t_1) and the agent in period 2 (t_2). Rational agents reason by backward induction. So, in period 1, the agent's reasoning (as t_1) will go something like this: I can either stay on the east side or go to the middle. If I go to the middle, then t_2 will have to choose whether to go on to the west side or return to the east side. Since I expect t_2 to want to be on the west side, I deduce that he would go on rather than back. Thus, since I want t_2 to get to the west side, I should go to the middle. t_1 accordingly, crosses to the middle of the road. Then, in period 2, t_2 notes that he is in the middle of the street and reasons: I would rather be on the west side than the east side, so I should go to the west side.

This type of reasoning gets the person to the other side of the street. However, it feels intuitively odd: the "I" in the reasoning refers to the transient time slice, not the person over time. There is an absence of any sense of agency over the whole period, of a continuing self who has interests that extend over the whole period and who can form an intention to cross the street and then just carry it out. In period 1, our agent can predict that he will continue the action in period 2. However, in neither period can he think of himself as performing the action of crossing the street; he cannot perceive himself as a continuing agent, nor act on his intentions qua continuing agent.

A standard decision theorist might counter that the agent would not usually model crossing the street in this manner. Decision theory is flexible about the length of time for which a transient agent exists. It is not committed to the fleeting time slices found in metaphysics. One would not usually model separate transient agents when the transient agents' interests are aligned because this situation does not usually present an interesting problem. There are two replies to this.

The first is that the time slice modeling strategy is not merely confined to situations of conflict of interests; people *have* modeled transient agents whose interests are aligned. Even when interests are aligned, there are interesting and perplexing issues. In particular, decision theory cannot necessarily predict that transient agents will coordinate over time in sequential coordination games (Binmore 1987; Pettit and Sugden 1989; Reny 1992; Gold and Sugden 2006).

The second response is that the decision theorist is still missing something interesting. The intrapersonal coordination example was supposed to show how the model does violence to our intuitions by not acknowledging agency over time. However, it is not critical to view intrapersonal coordination as a problem in order to think we need to supplement the standard model in the case of self-control. For the decision theorist, there is a problem of self-control when interests conflict, but no problem when they are aligned. The standard phenomenology of temptation involves feeling conflicted between long- and short-term interests, and a natural way to think of self-control is as the ability to align one's short- and long-term interests. Decision theory does not include this conflict, nor does it explain how some agents can resolve it without the use of external crutches. Standard decision theory has nothing to say about why preferences are sometimes aligned and other times not.

Decision theory provides a neat model of lack of self-control but has a lacuna when it comes to self-control. The crucial feature in the decision-theoretic model of self-control is a temporary and anticipated preference change. In order to exhibit self-control, an agent must bring her future preferences into line. In the model, this can only be done by changing the environment. What the model lacks is an internal mechanism by which an agent can bring her preferences into line or an explanation why, absent external mechanisms, one agent can exercise self-control in the face of temptation when another agent cannot.

10.4 Interpersonal Team Reasoning

While the transient agent model does very well at capturing the conflict of interests between time slices that can lead to problems of self-control, it does not capture our sense of agency over time or the role of willpower, in the sense described in philosophy and psychology. In order to capture agency over time and willpower without losing the insights of the transient

Table 10.1 *Hi-Lo Game*

		P_2	
		High	Low
P_1	High	2, 2	0, 0
	Low	0, 0	1, 1

agent model, we can introduce another level of agency, the person over time, as well as the level of the transient agents. In other words, we need a model of multiple levels of agency. Luckily, such a model already exists in the interpersonal case, which we can apply to our problem.

The theory of *team reasoning* was motivated by two families of game that have counterintuitive solutions (Colman and Gold 2017; Karpus and Gold 2016). One family of games has multiple Nash equilibria, but one of the equilibria *Pareto dominates* the others – there is one outcome in which all players are better off – yet game theory cannot recommend or predict the strategies that lead to the Pareto-dominant outcome. All classical game theory can say is that the rational solution will be one of the Nash equilibria. One member of this family is the coordination game known as Hi-Lo, shown in Table 10.1; another is the Stag Hunt. We can illustrate the problem using the Hi-Lo game. Standard game theory says that a rational player will choose a best reply to the other player(s). If P_1 chooses *high*, then the best response by P_2 is also to choose *high*, so (*high, high*) is a Nash equilibrium. However, if P_1 were to choose *low*, then the best responses by P_2 is to choose *low* as well; if P_1 chooses *low* and P_2 chooses *high*, then both get nothing. Therefore, (*low, low*) is also a Nash equilibrium. Standard game theory recommends that rational players play their parts in a Nash equilibrium, but it cannot advise one Nash equilibrium over another, so it cannot recommend to the players that they both play *high*.

The second family of games consists of those with a single Nash equilibrium that is Pareto dominated by a nonequilibrium outcome. In this case, game theory would recommend and predict that the strategies leading to the nonequilibrium outcome will *not* be played. An example of this type of game is the infamous prisoner's dilemma or its multiplayer version, the public goods game.

Team reasoning can explain why it is rational for individual players to choose the strategies that lead to the Pareto-dominant outcomes. The idea is that when an individual identifies with and reasons as a member of a team, he considers which *combination* of actions by members of the team would best promote the team's objective and then performs his part of that combination. Instead of asking, "What should *I* do?," as per classical game theory, players can ask, "What should *we* do, and how can I play my part?" It is clear that if there is common knowledge that all players group identify and are team reasoning, the theory of team reasoning can recommend and predict *high* play in Hi-Lo. In the prisoner's dilemma, if the off-diagonal outcomes are viewed as worse than the (C, C) outcomes from the perspective of the team, then with common knowledge of team reasoning the theory can predict and recommend C play. (For a more detailed explanation, see Gold and Sugden [2007a, 2007b].)

Team reasoning involves both a payoff transformation, to what Sugden (1993) calls "team-directed preferences," and an agency transformation, taking the relevant unit of agency to be the group. In behavioral economics, theorists often start with the material payoffs that subjects face and talk of their transformation into the utility payoffs that guide behavior, which may diverge from their material payoffs (for instance, if they care about what the other player gets). In the theory of team reasoning, we start with the utility payoffs that represent what the player wants to achieve as an individual, and when an individual group identifies, the payoff transformation is to the payoffs that the player wants to achieve as a team member (Gold 2012). Payoff transformation alone will not suffice; the agency transformation is a necessary part of the process. To see why, consider what would happen if we only had payoff transformation. No plausible payoff transformation will change the ordering of the payoffs in Hi-Lo, where interests are already aligned (see Karpus and Gold [2016] or Colman and Gold [2017] for a more extended explanation). In the prisoner's dilemma, payoff-transformation theories usually turn the (C, C) outcome into an equilibrium, but they do not change the equilibrium status of the (D, D) outcome, so they still cannot predict cooperative choices (Gold [2012] provides more detail). In order to see this, take the prisoner's dilemma on the left-hand side of Table 10.2. Transforming the game using golden-mean altruism, where each player is motivated to maximize the average of the player's outcomes, gives the matrix on the right-hand side of Table 10.2, which is a Hi-Lo matrix (see also Gold and Sugden [2007a, 2007b]).

Team reasoning was developed separately by Bacharach (1997, 1999, 2006) and Sugden (1993, 2000, 2003), and they have different explanations about when and why team reasoning occurs. Both Bacharach's and

Table 10.2 *Prisoner's Dilemma and Prisoner's Dilemma
Transformed*

		P_2				P_2	
		C	D			C	D
P_1	C	4, 4	0, 5	P_1	C	4, 4	2.5, 2.5
	D	5, 0	3, 3		D	2.5, 2.5	3, 3

Sugden's theories involve framing and expectations, but Bacharach's emphasis is on framing, whereas Sugden's is on expectations.

For Bacharach, team reasoning is a psychological process. Whether or not someone team reasons simply depends on whether she frames the game as a problem for "me" or a problem for "us." In an *unreliable team interaction*, there is some doubt as to whether other team members group identify and team reason. When deciding what to do, someone who team reasons will use *circumspect team reasoning*, taking into account the probability that other players team reason and maximizing expected utility from the perspective of the team. For instance, in the prisoner's dilemma, cooperating may not maximize the team utility if there is a large enough chance that the other player does not team reason, so circumspect team reasoning does not lead to unconditional cooperation. However, for Bacharach, team reasoning does not follow from rationally accessible deliberation, and team reasoning may leave the individual worse off in terms of her individual lights, for instance, if the team were to rank the off-diagonal (*C*, *D*) and (*D*, *C*) outcomes of the prisoner's dilemma higher than the (*C*, *C*) outcome or if circumspect team reasoning recommends that *C* play would maximize expected utility ex ante, but ex post the other player turns out not to have group identified.

For Sugden, team reasoning is a part of a social contract theory, where an individual can choose to cooperate with others for mutual advantage (Sugden 2011, 2015). If an individual sees that it is possible to frame a game as a problem for "us," then she may decide to team reason. However, no individual would team reason unless it furthered her individual interests, which puts constraints on the team payoff ordering. Sugden's team reasoners will not risk getting suckered, which also means that they will not team reason without assurance that others are team reasoning too; hence we can call it *mutually assured team reasoning*. The idea that people can decide to team reason, the constraints on the team preferences, and the need for assurance are all points of difference with Bacharach's theory (see Gold [2012] for a more detailed comparison).

10.5   Intrapersonal Team Reasoning

Problems of self-control are problems of intrapersonal cooperation (Gold 2013). The classic interpersonal problem of cooperation is the prisoner's dilemma, a type of *public goods game* where costly actions by individuals have positive externalities, so it is individually rational for an individual to defect (not to contribute to the public good), even though all individuals prefer the situation where everyone contributes to the one where no one contributes. In other words, one agent takes an action whose benefits are spread across many agents. The benefit that accrues to the individual does not outweigh the cost of the action, but the benefit that accrues across all individuals does. Problems of self-control are similar in that they involve one transient agent paying a cost in return for benefits that accrue to other transient agents, so they are problems of intrapersonal cooperation.

By analogy, if interpersonal team reasoning can lead to cooperation in the prisoner's dilemma, then intrapersonal team reasoning can promote self-control. In the intrapersonal case, the team consists of the set of time slices that make up the person over time. The units of agency are the time slice and the self over time, and the equivalent of group identification is identifying with the person over time. In the standard model, the time slice does "transient-agent reasoning," asking, "What should I now do?," whereas intrapersonal team reasoning allows any time slice that identifies with the person over time to ask, "What should the I the person over time do?" and to play its part in the best team plan.

Take the problem of Jo's examination, introduced by Gold and Sugden (2006). This is a three-period model with three transient agents, Jo_1, Jo_2, and Jo_3. In period 3, Jo takes an examination; in periods 1 and 2, Jo must decide whether to study for the exam or to relax. The experienced utility (in the sense of Kahneman and Thaler 2006) of studying in any period is −3, whereas that of resting is 0. In period 3, experienced utility is 0 if Jo has rested on both previous days, 5 if she has rested on one day and studied on the other, and 10 if she has studied on both days. In terms of experienced utility, if either Jo_1 or Jo_2 chooses to study, then that time slice bears a cost that has benefits for Jo_3. The benefits of studying are greater over the lifetime than the costs. However, the transient agents that study do not capture the benefits. Table 10.3 shows the payoffs for each combination of moves in terms of experienced utility.

Table 10.3 *Jo's Examination: Experienced Utility*
Payoffs for Jo_1, Jo_2, and Jo_3

		Jo_2	
		Study	Rest
Jo_1	Study	-3, -3, 10	-3, 0, 5
	Rest	-3, 0, 5	0, 0, 0

Table 10.4 *Jo's Examination: Payoffs in Lifetime*
Utility for Jo_1, Jo_2, and Jo_3

		Jo_2	
		Study	Rest
Jo_1	Study	1, 1, 14	-1, 2, 7
	Rest	2, -1, 7	0, 0, 0

Note: Lifetime utility for player i is the sum of the experienced utilities of all other players plus two times the experienced utility of player i, representing the time slice's double weighting of its own outcomes.

Even if each transient agent cares about the others, a little bit of present bias can still lead to a problem of self-control. Imagine that as in the O'Donaghue and Rabin (1999) model earlier, each transient agent values the experiences of all transient agents but places double the weight on its own experiences as on those of other time slices. Table 10.4 shows the payoffs in terms of these preferences over the lifetime. Now some of the benefits of studying accrue to the transient agent who studies (because the payoffs of the other transient agents are in her utility function), but there is still an externality, and the costs of studying still outweigh the benefits for each individual transient agent. In this lifetime preference model, Jo_1 and Jo_2 are playing a sequential prisoner's dilemma. The dominant strategy is *rest*, but every transient agent prefers the outcome of (*study, study*) to those of (*rest, rest*). By backward-induction reasoning, Jo_1 can predict that Jo_2 will choose *rest*. So Jo_1's choice is effectively between the sequences (*study, rest*) and (*rest, rest*); (*rest, rest*) is superior from her point of view, so according to decision theory she should choose *rest*.

However, if we allow that each transient agent can ask, "What should *I* the person over time do?," then it may be possible to achieve the outcome (*study, study*). Intrapersonal team reasoning can solve the intrapersonal problem of cooperation in the same way that interpersonal team reasoning solves the interpersonal problem. Intrapersonal team reasoning could proceed according to either of Sugden's or Bacharach's theories.

In the game in terms of lifetime utilities (Table 10.4), there is an opportunity for mutual benefit, so we can apply Sugden's mutually assured team reasoning. If Jo_1 has reason to believe that Jo_2 will identify with the team of the person over time and will endorse mutually assured intrapersonal team reasoning, then if Jo_1 also endorses mutually assured intrapersonal team reasoning, she can choose to *study*. This captures our intuition that starting a plan that will require a series of sacrifices, such as a study plan or a diet, requires the belief that our later self will follow through.

However, we might wonder whether mutually assured team reasoning is the right framework for thinking about the self over time. It is built on ideas of reciprocity and the social contract, which do not seem to apply in the case of the self over time. The lifetime utility is constructed from the time slices' preferences, with the transient agents compromising on their time slice preference satisfaction. As well as the basic implausibility of this approach, we might also worry that it introduces an element of double counting into the goals of the person over time. In the realm of social contract theory, Dworkin (1977) distinguished "personal preferences," which are wholly about oneself, and "external preferences," which are about other people. He argues that people's external preferences should not influence the assignment of goods. In the intrapersonal case, if Jo_1 is positively disposed toward Jo_2 and wants her to have good outcomes and, as well as allowing Jo_2's personal preferences to influence what the team seeks to achieve we also take into account Jo_1's external preference, then we have double counted Jo_2's outcomes.

We can also apply Bacharach's circumspect team reasoning because Jo_1 has to make her choice before she knows for sure whether Jo_2 will group identify and team reason. In the Bacharach framework, there is no particular reason to think that the lifetime utility function is based on each transient agent's lifetime preferences rather than on each transient agent's outcomes. As a simplification, let us assume that the lifetime function is achieved by aggregating the transient agents' experienced utility. However, in that case, as this stands, the model would lead to unconditional

Table 10.5 *Jo's Examination Threshold Case:*
Experienced Utility Payoffs for Jo₁, Jo₂, and Jo₃

		Jo_2	
		Study	Rest
Jo_1	Study	-3, -3, 10	-3, 0, 0
	Rest	0, -3, 0	0, 0, 0

cooperation (self-control) by any time slice who team reasons. This is guaranteed by the externality structure of the problem: the cost borne by the time slice is always outweighed by the benefits to the set of time slices, so *study* will always be better for the team regardless of whether the other time slices group identify. In fact, we might even wonder why we need team reasoning. If a time slice simply takes the aggregated utility of all time slices as their end, this would suffice to get them to exercise self-control. So this model both violates the intuition that a person will not usually start on a plan unless he expects his later selves to follow through and obviates the need for team reasoning.

We can reintroduce an element of conditional cooperation and, with it, a need for team reasoning if we turn the examination problem into a threshold case. A *threshold public good* does not have a linear relationship between costs and benefits. Rather, the good is provided if and only if contributions pass a minimum threshold. As applied to the problem of Jo's examination, imagine that the exam is pass/fail and Jo needs to study both days in order to pass. The only change we need make to the original problem is in period 3, where experienced utility is 0 unless Jo has worked on both previous days, in which case it is 10. Now the outcome matrix in experienced utilities is as in Table 10.5, and the aggregated outcomes when both players view the problem from the perspective of the intrapersonal team are as in Table 10.6. From the perspective of the team, this is a Hi-Lo game. We can see that (*study, study*) gives better outcomes for the team than (*rest, rest*), but either of these is better than the outcome where one time slice works and the other rests. Therefore, if Jo₁ group identifies and team reasons, then what she should do depends on whether or not she expects Jo₂ to group identify. If she expects that Jo₂ will also team reason, then she should choose *study*, but if she expects that Jo₂ will not team reason and will therefore choose the individually dominant strategy of *rest*, then Jo₁ should *rest* herself. Whether or not a team-reasoning transient

Table 10.6 *Jo's Examination Threshold Case: When
Both Players View the Problem from the Perspective
of the Intrapersonal Team and Aggregate the
Transient Agents' Experienced Utilities to Obtain the
Team Payoffs*

		Jo_2	
		Study	Rest
Jo_1	Study	4, 4, 4	−3, −3, −3
	Rest	−3, −3, −3	0, 0, 0

agent will exercise self-control will depend on the payoffs involved and on the strength of her belief that later time slices will also team reason.

It is not difficult to work out when a team-reasoning Jo_1 will choose *study*. We know that if Jo_1 does not study, then the best response by Jo_2 will be *rest* from the perspectives of both the time slice and the team over time: choosing *rest* is the unconditional best response from the perspective of the time slice, and it is the best response for a team-reasoning Jo_2 if Jo_1 has chosen *rest*. Remembering that a team-reasoning Jo_1 will maximize the payoff of the team and assigning $0 < p < 1$ as the probability that Jo_2 will group identify and U_t as the team payoff function, then Jo_2 will *study* if

Expected team payoff if Jo_1 chooses *study* > expected team payoff if Jo_1 chooses *rest*

$$=> [p(Jo_2 \text{ chooses } study) \times U_t(study, study)] + [p(Jo_2 \text{ chooses } rest) \times U_t(study, rest)] > U_t(rest, rest)$$

$$=> 4p - 3(1 - p) > 0$$

$$=> 7p > 3$$

$$=> p > 3/7$$

Therefore, Bacharach-style intrapersonal team reasoning, understood as a psychological process of identifying with the person over time, can generate a plausible theory of rational self-control (one that is conditional on what later time slices are expected to do) if the structure of the intrapersonal problem is a threshold public goods game. It is not implausible to think that problems of self-control, as viewed by the person involved, have this threshold structure. Although most self-control problems have an underlying continuous public goods game structure, we may have a tendency to turn them into threshold cases. Most self-control problems have continuous but imperceptible benefits. Think of smoking, where every cigarette has a very small negative effect on the smoker's health, or dieting, where every calorie that the dieter forgoes consuming puts her nearer to fitting into that dress. However, when we forgo these temptations, we are looking for perceptible benefits. The smoker wants to feel healthier; the dieter wants to lose a pound or to fit into a dress. These perceptible benefits fix a threshold – albeit one that is vague – the number of cigarettes or calories forgone to make a perceptible difference. If the person is aiming for a perceptible difference, then there is no point in an earlier self forgoing the first cigarette or the first dessert unless she expects enough of the later time slices to continue the good work.

Of course, the theory of intrapersonal team reasoning needs to be supplemented with an account of how time slices come to identify with the person over time. This is not the place to develop one, but here is a sketch. (I say more about it elsewhere, in Gold unpublished; Gold and Kyratsous 2017). Again, we can make an analogy to the interpersonal case. In psychology, there is a body of research about how individuals come to identify with groups. The mechanisms of group identification fit into two broad categories: recognizing that the group members have some sort of shared goal or common fate and recognizing commonalities or similarities between the individuals within the group. Both of these could apply to the intrapersonal case.

Time slices may identify with the person over time because they recognize that they all share long-term interests. As Korsgaard (1989) argues, there is a sense in which time slices are one continuing person *because* they have one life to lead. Her arguments are normative, but a psychological and phenomenological analogue can be found in the work of James (1890), where one source of a sense of self over time is the recognition that a self in the past or the future was or will be part of the same person. Therefore, we might speculate that increasing the salience of the shared interests of the time slices, or their long-term goals, will facilitate this sense of identification.

Alternatively, time slices might identify with the person over time because they realize that they are either similar to or connected to the other time slices. James (1890) also thought that the current self's perception that it is similar to proximate selves gives rise to a sense that it is continuous with those proximate selves. Psychologists have found that subjects who rated themselves as more connected to later selves, in the sense of Parfit (1984), were more patient (Bartels and Rips 2010) and that connectedness can be manipulated, resulting in increased patience (Bartels and Urminsky 2011). Accordingly, increased salience of either similarities or connectedness between the time slices may facilitate identification with the person over time.

Decision theory provides a model of instrumental rationality in which decision makers take the best means to their ends. Instrumental rationality presupposes that the decision maker has a set of ends. Therefore, the time slice has to identify with a level of agency and take on a set of ends before instrumental reasoning can begin. However, decision theory has nothing to say about phenomenology. It is consistent with this picture that the time slice experiences a tension between the transient-agent preferences and the self-over-time preferences, so the model is compatible with the phenomenology of conflict.

10.6 Willpower, Decision Theory, and Intentions

Intrapersonal team reasoning sheds fresh light on willpower. In the model, willpower is the ability to align one's present self with one's extended interests by identifying with one's self over time. This picture of willpower differs from the idea of Holton (1999, 2009) that strength of will consists of not reconsidering one's resolutions. But it does create a space for resolutions in decision theory and resolves a puzzle about resolutions that we find in Holton's account.

Standard decision theory does not have room for intentions – understood as motivating plans – or for resolutions that are designed to fortify us against contrary inclinations later on. In O'Donaghue and Rabin's (1999) model, the naive agent who forms a plan in t_0 to write the report in t_1 will not carry out the plan and ends up missing the Depp movie. Part of her problem was that she did not take into account the preferences that her t_1 self would have. A sophisticated agent would correctly predict her future present bias, and her t_0 self would be able to plan to do the report in t_2. However, this is simply a correct prediction of her future behavior. If prediction is all that planning consists of, then we do not need a separate concept of a "plan," and there is no need for intentions because we can suffice with beliefs. Further, the hyperrational agents of standard economics can make optimal decisions in a flash; they can make them whenever and however many times as they want, so there is no need to make them in advance and form the intention to act on the decision later.

Decision theory can make room for intentions by appealing to the idea of bounded rationality. If it takes time for a boundedly rational agent to make a decision, then it may not be best for the agent to take the decision at the time of action. Once the person has made a decision, then, other things being equal, she should not waste her limited time by reopening the question. Hence it might be optimal for the agent to make the decision in advance and form an intention as a reminder of her decision, which she can consult at the time of action.

However, the idea of a resolution fits uncomfortably in this framework. Remember that agency sits with time slices. Effectively, the past time slice takes a decision, and, other things being equal, the future time slice does not reopen the question. (If decision making is onerous, then there may also be a problem of procrastination about making the decision, but we leave that aside here.) One part of "other things being equal" is the idea that the later time slice would be likely to make the same decision as previously, so redeliberating is a waste of time. This seems uncontroversial in cases where there is no conflict of interests between time slices. However, Holton's (1999, 2009) resolutions are formed in order to defeat contrary inclinations, which will arise at the time of action. In the decision-theoretic framework, the past time slice is making a decision that she knows the future time slice will not want to carry out if the future time slice takes the standard time slice perspective. Therefore, if a time slice remembers that an earlier time slice made a resolution, she also has reason to think that the resolution conflicts with her current time slice preferences, so if she is thinking as a time slice, then she should abandon any prior resolutions. (This relates to Hinchman's [2003] idea that diachronic agency involves a type of self-trust.)

If the time slice is to act on resolutions, then we need to add something extra, and intrapersonal team reasoning can supply the missing piece of the puzzle. In the framework of the person as a team over time, an intention is a plan made by an earlier time slice who identifies as a member of the team over time. The intention has two different purposes. First, it is a contingency plan for later time slices who turn out to identify with the person over time and therefore share the team preferences of the planner. There is no conflict of interest, and if the later time slice has no reason to suspect that the earlier planning time slice made a bad plan, then she can simply follow her part of the plan. Intrapersonal team reasoning can explain how the different time slices' interests are aligned so that the later time slice knows that she should follow the plan made by the earlier one.

Planning can also play a second type of role in this picture. In the same way that standard decision theory allows earlier time slices to take actions to constrain later time slices, in the theory of intrapersonal team reasoning, the earlier time slice can take actions that increase the probability of group identification by later ones. Remembering a plan may encourage the time slice that does the remembering to identify with the person over time. For instance, it makes salient the existence of the temporally extended agent and the shared extended interests of the time slices.

In this second role, making a plan may have some of the effects of Holton's (1999, 2009) resolutions. By encouraging the later time slice to identify with the person over time and therefore to act on the plan, it may prevent the transient agent reasoning that leads to weakness of will. Therefore, resolutions are a mechanism of self-control. Nevertheless, in the model of intrapersonal team reasoning, the resolution is not the root cause of self-control. The fact that the plan can be effective in the face of contrary inclinations is parasitic on the idea that the time slice can identify with the self over time and do intrapersonal team reasoning. If, at the time of action, the time slice did some reasoning, then she could come to the same decision as previously, provided that she does intrapersonal team reasoning rather than transient agent reasoning. Further, an agent who makes a resolution but then happens to reconsider at time of action is not totally lost (so intrapersonal team reasoning solves a problem posed by Bratman [2014] about how an agent can rationally form an intention if he anticipates that he will reopen the question and take a transient agent view at the time of action). It is not a foregone conclusion that the later time slice will do transient agent reasoning rather than intrapersonal team reasoning.

We can also compare willpower and resolutions to willpower as intrapersonal team reasoning using the philosophical framework of synchronic versus diachronic self-control (Mele 1987). *Diachronic self-control* occurs when an agent anticipates a preference change and takes action to prevent himself from succumbing later. *Synchronic self-control* occurs when the agent exercises self-control at the very same time as experiencing a temptation. Intentions can be a means of diachronic self-control in both the resolution and the intrapersonal team-reasoning accounts. In the resolutions account, this is because not reconsidering one's resolution is the instrument of synchronic self-control. In the intrapersonal team-reasoning account, an agent who identifies with the person over time might not reconsider his intentions. However, the ultimate instrument of synchronic self-control, which also underpins the intention or resolution, is intrapersonal team reasoning. If an agent forms a resolution, it is effective because it prompts identification with the person over time and, hence, acting on the results of intrapersonal team reasoning.

10.7 Conclusion

I have presented a picture of willpower as intrapersonal team reasoning, analogous to using interpersonal team reasoning to solve problems of cooperation between individuals. I suggested that we should model problems of self-control as threshold public goods games. I have shown how, although the time slices' transient agent preferences are in conflict, it is possible for them to identify with the person over time and use circumspect intrapersonal team reasoning to resolve their problem of self-control. Intrapersonal team reasoning also provides a basis for introducing intentions and resolutions into decision theory, although at base it is intrapersonal team reasoning that solves the synchronic problem of self-control and that gives resolutions their power.

I have shown how willpower can be instrumentally rational for the person over time, even though succumbing to temptation is instrumentally rational from the perspective of the time slice. In this sense, the model provides an answer to the long-standing philosophical question of how an individual can intentionally act against what she judges to be best. Many people have the intuition that the time slice is doing something wrong if she succumbs to temptation. It follows from what I have said here that the wrongness is not derived from instrumental rationality. There is a lot more to be said about why time slices should identify with the self over time and when they will do so (Gold unpublished). But this is an issue for a whole separate chapter.

References

Ainslie, G. 1992. *Picoeconomics*. Cambridge: Cambridge University Press.

Bacharach, M. 1997. *"We" Equilibria: A Variable Frame Theory of Cooperation.* Oxford: Institute of Economics and Statistics, University of Oxford.

Bacharach, M. 1999. Interactive team reasoning: a contribution to the theory of co-operation. *Research in Economics* 53:117–47.

Bacharach, M. 2006. *Beyond Individual Choice: Teams and Frames in Game Theory*. Princeton, NJ: Princeton University Press.

Bartels, D. M., and L. J. Rips. 2010. Psychological connectedness and intertemporal choice. *Journal of Experimental Psychology: General* 139(1):49–69.

Bartels, D. M., and O. Urminsky. 2011. On intertemporal selfishness: how the perceived instability of identity underlies impatient consumption. *Journal of Consumer Research* 38(1):182–98.

Binmore, K. 1987. Modeling rational players, part I. *Economics and Philosophy* 3:9–55.

Bratman, M. E. 2014. Temptation and the agent's standpoint. *Inquiry* 57 (3):293–310.

Colman, A. M., and N. Gold. 2017. Team reasoning: solving the puzzle of coordination. *Psychonomic Bulletin & Review* 2017:1–14.

Dworkin, R. 1977. *Taking Rights Seriously* (vol. 136). Cambridge, MA: Harvard University Press.

Gold, N. 2012. Team reasoning, framing and cooperation. In *Evolution and Rationality: Decisions, Co-operation and Strategic Behaviour*, ed. S. Okasha and K. Binmore (chap. 9, pp. 185–212). Cambridge: Cambridge University Press.

Gold, N. 2013. Team reasoning, framing, and self-control: an Aristotelian account. In *Addiction and Self-Control: Perspectives from Philosophy, Psychology, and Neuroscience*, ed. N. Levy. Oxford: Oxford University Press.

Gold, N. Unpublished manuscript. Guard against Temptation.

Gold, N., and M. Kyratsous. 2017. Self and identity in borderline personality disorder: agency and mental time travel. *Journal of Evaluation in Clinical Practice* 23(5):1020–28.

Gold, N., and R. Sugden. 2006. Conclusion. In *Beyond Individual Choice*, ed. M. Bacharach. Princeton, NJ: Princeton University Press.

Gold, N., and R. Sugden. 2007a. Collective intentions and team agency. *Journal of Philosophy* 104:109–37.

Gold, N., and R. Sugden. 2007b. Theories of team agency. In *Rationality and Commitment*, ed. F. Peter and H. B. Schmid. Oxford: Oxford University Press: 280–312.

Greer, J. M., and D. K. Levine. 2006. A dual-self model of impulse control. *American Economic Review* 96(5):1449–76.

Hinchman, E. S. 2003. Trust and diachronic agency. *Noûs* 37(1):25–51.

Holton, R. 1999. Intention and weakness of will. *Journal of Philosophy* 96(5): 241–62.

Holton, R. 2009. *Willing, Wanting, Waiting.* Oxford: Oxford University Press.

James, W. 1890. *The Principles of Psychology.* New York, NY: H. Holt and Company.

Kahneman, D., and R. H. Thaler. 2006. Anomalies: utility maximization and experienced utility. *Journal of Economic Perspectives* 20(1):221–34.

Karpus, J., and N. Gold. 2016. Team reasoning: theory and evidence. In *Handbook of Philosophy of the Social Mind*, ed. J. Kiverstein (pp. 400–17). New York, NY: Routledge.

Korsgaard, C. M. 1989. Personal identity and the unity of agency: a Kantian response to Parfit. *Philosophy & Public Affairs* 1989:101–32.

Loewenstein, G., T. O'Donoghue, and M. Rabin. 2003. Projection bias in predicting future utility. *Quarterly Journal of Economics* 118:1209–48.

Lowenthal, D., and G. Loewenstein. 2001. Can voters predict changes in their own attitudes? *Political Psychology* 22(1):65–87.

Mele, A. R. 1987. *Irrationality: An Essay on Akrasia, Self-Deception, and Self-Control.* Oxford: Oxford University Press.

Mele, A. R. 1992. *Irrationality: An Essay on Akrasia, Self-Deception, and Self-Control.* Oxford: Oxford University Press.

O'Donoghue, T., and M. Rabin. 1999. Doing it now or later. *American Economic Review* 89(1):103–24.

Parfit, D. 1984. *Reasons and Persons.* Oxford: Oxford University Press.

Pettit, P., and R. Sugden. 1989. The backward induction paradox. *Journal of Philosophy* 86:169–82.

Pronin, E., and L. Ross. 2006. Temporal differences in trait self-ascription: when the self is seen as an other. *Journal of Personality and Social Psychology* 90 (2):197–209.

Reny, P. 1992. Backward induction, normal form perfection and explicable equilibria. *Econometrica* 60:627–49.

Schelling, T. C. 1984. Self-command in practice, in policy, and in a theory of rational choice. *American Economic Review* 74(2):1–11.

Sugden, R. 1993. Thinking as a team: towards an explanation of nonselfish behavior. *Social Philosophy and Policy* 10:69–89.

Sugden, R. 2000. Team preferences. *Economics and Philosophy* 16:175–204.

Sugden, R. 2003. The logic of team reasoning. *Philosophical Explorations: An International Journal for the Philosophy of Mind and Action* 6:165–81.

Sugden, R. 2011. Mutual advantage, conventions and team reasoning. *International Review of Economics* 58:9–20.

Sugden, R. 2015. Team reasoning and intentional cooperation for mutual benefit. *Journal of Social Ontology* 1:143–66.

Strotz, R. H. 1955–56. Myopia and inconsistency in dynamic utility maximization. *Review of Economic Studies* 23:165–80.

Thaler, R. H., and H. M. Shefrin. 1981. An economic theory of self-control. *Journal of Political Economy* 89(2):392–406.

Van Boven, L., G. Loewenstein, and D. Dunning. 2005. The illusion of courage in social prediction: underestimating the impact of fear of embarrassment on other people. *Organizational Behavior and Human Decision Processes* 96(2):130–41.

CHAPTER 11

The Many Ways to Achieve Diachronic Unity

Kenny Easwaran and Reuben Stern

11.1 Introduction

When Zeb's owner, Martin, placed his favorite toy in front of him, Zeb did something that surprised Martin. Rather than immediately rushing to put the toy in his mouth, as Martin expected, Zeb spent about 15 seconds concentrating his efforts on *not* putting the toy in his mouth, before appearing to allow himself the opportunity to feverishly pounce on the toy.

Martin was very impressed. As he saw things, Zeb's behavior seemed to exhibit some distinctly human property – i.e., some property that we typically associate with human agency but not animal agency. What exactly this property was, or whether it was definitely present, Martin could not say. But he was sure that it had something to do with Zeb's apparent ability to execute a plan over time.

Although Martin may not have been able to articulate why he found Zeb's behavior impressive, Bratman (1987) has argued that Zeb's seeming capacity to self-govern over time is suggestive of a capacity to form binding future-directed intentions. As Bratman sees things, if there are *time slices* of Zeb that would prefer to pounce on the toy but that nevertheless abstain, then they must do so because they are constrained or governed by some prior future-directed intention of Zeb's. In order for an agent's actions to be bound in this way, there must be rational pressure to follow through with one's prior plans. But it is no small task to determine exactly when, or in what way, agents are bound by their prior intentions because there are clearly contexts in which it is rationally permissible to abandon some prior future-directed intention – e.g., when Zeb initially intends to abstain for 15 seconds but subsequently learns that Martin will take the toy away after 10 seconds.

Bratman (2012) argues that the bindingness of future-directed intentions is rooted in a diachronic norm of coherence that rules out certain combinations of intentions across time. Roughly, Bratman argues that agents must stick with their prior intentions or plans unless they acquire reason to abandon them. So, according to Bratman, unless Zeb acquires a reason to abandon his prior plan to abstain from pouncing for 15 seconds, then he must stick with his prior intention on pain of diachronic incoherence.[1]

In this chapter, we first argue that the diachronic norm of coherence that Bratman defends is false, and then we identify circumstances in which agents can successfully bind their subsequent time slices without any recourse to future-directed intentions or diachronic norms of coherence. Thus we argue that there are means through which to achieve diachronic unity or self-governance that do not rely on the cognitive resources that Bratman attributes to planning agents.[2] This does not mean, however, that we think that there is no role for binding future-directed intentions in diachronic agency. Rather, we show that in some of the circumstances where agents can achieve diachronic unity through alternative means, it is still true that agents would be more unified and better off (in expectation) were their later actions to be governed by their prior intentions. Thus we argue that future-directed intentions play a valuable, but not essential, role in unifying diachronic agents across time.

II.2 Bratmanian Self-Governance

According to Bratman (2012: 79), planning agents have reason to treat their prior intentions as defaults when doing so follows from adherence to the following diachronic norm D:

> D: The following is locally irrational: intending at t_1 to x at t_2, throughout t_1–t_2 confidently taking one's relevant grounds adequately to support this very intention, and yet at t_2 newly abandoning this intention to x at t_2.[3]

Bratman's defense of D is twofold. First, he argues that D is initially plausible insofar as it yields intuitive verdicts when applied to cases. Second, Bratman argues that self-governance over time is possible only given conformity to D. In this section, we argue that Bratman is wrong on both counts.

---

[1] Bratman would depict Zeb as acquiring reason to abandon his prior intention on learning that Martin will take the toy after ten seconds.

[2] When we speak in terms of "achieving diachronic unity," we do not mean to take a stance on the difficult question of what precisely constitutes diachronic unity. If some readers are unhappy with our choice of vocabulary because they think that diachronic unity cannot be understood behaviorally, they should feel free to substitute every mention of the term with its closest behavioral relative.

[3] Bratman clarifies what he means by "locally irrational" in the following sentence: "*D* states a *local* demand that concerns a specific sub-cluster of attitudes within an overall cross-temporal psychic economy" (2012: 79, emphasis added).

As Bratman realizes, a reasonable first reaction to D is that even if true, D never compels an agent to act in accordance with her previous intention because the circumstances in which D applies are circumstances in which the agent already takes her grounds to support acting in accordance with her intention.[4] That is, since D bars the agent from abandoning her prior intention only when she already "takes her relevant grounds to adequately support" her prior intention, and since these are circumstances in which it is already settled what the agent should rationally do, it may seem that the class of cases where D accounts for the *governance* of future-directed intentions is trivially small (or empty).[5]

If an agent's grounds could only ever adequately support one of her options, then D would, in fact, collapse into triviality (because abandoning the original adequately supported intention for some alternative would necessarily consist of adopting a new intention that is, by the agent's lights, less optimal than the original). But since, as Bratman (2012: 79–80) points out, an agent's grounds can adequately support *multiple* incompatible alternatives – e.g., when an agent's grounds equally support two incompatible options or when the grounds that support two options are incommensurable – D does sometimes compel agents to act in accordance with their prior intentions. Consider D's application to the following case.

When Carmen woke up this morning, she settled on having an apple for her midafternoon snack. Then, when she arrived at her favorite fruit stand at 3 o'clock, she reconsidered and opted for an orange. Was Carmen rational?

There are clearly further specifications of this case where D does not compel Carmen. For example, if Carmen discovers that oranges are on sale or that the apples in today's selection are rotten, then Carmen plausibly no longer takes the relevant grounds to adequately support having an apple, and D therefore says nothing about whether Carmen has done something irrational. But there are also clearly further specifications of this case where D *does* compel Carmen. If Carmen initially takes the apple and orange to be equivalently good, or if she judges her choice to be, well, between "apples and oranges," and if she acquires no reason to prefer the orange

---

[4] One might reasonably point out that D never *compels* an agent because D is not a command or imperative but rather an evaluative standard. When we speak in terms of D "compelling" an agent to do something, we just mean that D says that the agent must do the thing in question in order to qualify as rational.

[5] Or as Bratman might put things, if the agent intends to x at t_1 and then at t_2 still takes her grounds to support x-ing, it may seem that there is no work for the prior intention to do in constraining what is rational, since the agent's assessment of the relevant grounds at t_2 are in agreement with what she intended at t_1.

to the apple before 3 o'clock (perhaps because she learns nothing new about the fruits), then *D* says that Carmen must stick with her original choice of the apple in order to be rational.

So far, so good. It is clear that Carmen should *not* be bound by her previous choice of the apple when she learns that the apples are rotten, and it is prima facie plausible that Carmen should stick with her prior intention if she acquires no new information that suggests that she should do otherwise. So it may seem that Bratman is right to champion *D* and that future-directed intentions govern by establishing defaults that determine what agents should do in the absence of new reasons.

Attractive as *D* may seem, we argue that it suffers from two main problems. First, although we agree with Bratman that an agent's grounds can adequately support multiple incompatible alternatives, we argue that the class of cases where *D* compels agents to act in accordance with their prior intentions is too small to account for the many cases where our current selves are rationally governed by our prior selves. Second, we argue that *D* is false – i.e., that adherence to *D* sometimes results in irrational behavior in the class of cases where *D* endows intentions with binding force.

Although Bratman is right that the class of cases where *D* construes intentions as governing is nonempty, it is relatively small. Suppose, for example, that Carmen acquires the slightest piece of information that speaks in favor of oranges over apples – say that oranges (but not apples) are within arm's reach on arriving at the fruit stand. In this case, *D* ceases to say anything about what Carmen should do because she no longer takes her relevant grounds to support opting for the apple. Alternatively, consider a variant of Carmen's case that more closely resembles the classic tale of Buridan's ass, in which Carmen finds herself positioned exactly between the apples and oranges in the grocery store. Even if Carmen initially takes her grounds to adequately support both kinds of fruit, she plausibly ceases to take her grounds to support opting for the orange once she takes even one step toward the apple (because the apple is then closer than the orange). Thus, here, just as in cases where the agent's grounds initially support only one option, *D* does not depict the agent's intention as governing because there is no tie to be broken.[6]

---

[6] In this section, we intend for *tie* to refer not only to cases where options are regarded as equally good but also to cases where options cannot be ranked. If Chang (2002) is right that there are cases of "parity," in which neither of two options is better than the other, but also not equally good, then these may be cases in which a small improvement to one of the two options does not make it better than the other. If true, the set of cases in which *D* applies is not trivially small. But it is still too small,

Since Carmen's prior choice seems to play a role in determining what is rational even in cases such as these – i.e., when Carmen acquires negligible reason to break from her plans or when Carmen acquires reason to follow through with her plans – it is hard to see why Bratman thinks that the truth of D can explain what he takes to be the essential role that future-directed intentions (or plans) play in self-governance over time.

Perhaps Bratman's thought is that in the many circumstances where an agent *has* acquired new information that changes what she takes her relevant grounds to support, her prior intention plays a governing role insofar as it is still true that *were* she not to have acquired this new information, then she *would* have been bound by her initial intention. If this is right, it is still hard to see why Bratman thinks that such a thin modal property can drive self-governance over time. When an agent self-governs over time by settling on some course of action, it seems that her governance consists of *actually* constraining the set of actions that the agent's subsequent time slices can rationally perform, *not* in constraining the set of actions that the agent's subsequent counterpart time slices can rationally perform in distant possible worlds. So, even if Bratman were right that there are not counterexamples to D, it is hard to see how D could play the central role in rational self-governance that Bratman attributes to it.

But, alas, even though D does not stick its neck out much, it sticks its neck out too much. That is, even within the range of cases in which D depicts intentions as governing, there are counterexamples.

Suppose that Carmen arrives at the fruit stand five minutes early and has time to kill before her friend meets her there at 3 P.M. Because there is little to do while standing at a fruit stand, Carmen reopens the question of which fruit to have in order to see what fruit she currently prefers. (It helps pass the time.) On doing so, she takes her grounds to adequately support both the apple and the orange, just as she did when she woke up this morning. But this time, she opts for the orange instead of the apple.

According to D, Carmen's behavior is irrational because she (1) intends at t_1 to have the apple at t_2, (2) takes her grounds to adequately support having the apple throughout t_1–t_2, and (3) abandons the intention to have the apple at t_2. But this seems wrong. Surely, there is nothing wrong with introspecting to see which fruit she currently wants given, first, that it is sometimes rationally permissible to reopen settled practical questions

we contend, for D to account for many cases in which our current selves are rationally governed by our prior selves.

and, second, that Carmen is better off for having done so in this case (because it helps her pass the time). But once Carmen checks what she currently wants, she does not have any reason to stick with her prior choice. Indeed, it seems that in order for Carmen to derive any benefit from reopening the question, she must regard her choice as unconstrained by her previous choice. This means that it seems rationally permissible for Carmen to plump for the orange at 2:55 for the very same reasons that it seemed rationally permissible to plump for the apple in the morning.[7]

Of course, the circumstances of Carmen's case are somewhat rare. By Carmen's lights, she is better off deliberating once more, but we usually think that there are costs (rather than benefits) to deliberating because doing so takes time and effort. This may prima facie suggest that cases like Carmen's should be set aside, but this is wrong. By focusing on these cases, it becomes apparent that whether an agent should reopen a question depends just on whether the agent is better off (in expectation) if she does so. Because we happen to live in a world in which there are usually costs (rather than benefits) to reopening questions, our prior choices actually do play an integral role in unifying agents across time through self-governance. But, if we lived in a world in which we were always bored, and in which we had cognitive resources to go around, our prior choices would not play this particular organizing role.

This suggests that when agents are bound by their previous choices depends just on considerations of expected utility – i.e., Carmen should stick to her prior plan when doing so makes her better off (in expectation) but otherwise should not. In Section 11.5, we address Carmen's case in some detail and propose a model according to which the (ir)rationality of switching plans depends on the costs and benefits of introspecting. But before developing this model, it is worth demonstrating that agents can achieve diachronic unity (and diachronic self-governance) even when there are not costs to introspecting and absent diachronic norms of coherence. Sections 11.3 and 11.4 describe two kinds of cases leading to diachronic unity absent any costs to introspection, and then Section 11.5 incorporates those costs.

All the cases we describe involve nearly maximally disunified selves composed of time slices that each have their own interests.

---

[7] One might contend that Carmen is irrational because she *reconsiders* her prior intention, not because of what she elects to do on reconsidering her prior intention. By our lights, this does not square with Bratman's own treatment of *D* because it is clear that *D* does not bar Carmen from reconsidering and sticking with the apple on doing so. Moreover, it seems rationally permissible for Carmen to reconsider since doing so alleviates her boredom.

We demonstrate three different features of preference structure that can yield some sort of diachronic unity even in these conditions and show that when multiple features are present, an even greater degree of unity occurs. While none of these models is very realistic, we think that they exemplify effects that are present to some degree in actual people in many situations. There are many types of preference structures that we have not investigated yet, such as cases where time slices care distinctively about their own behavior rather than that of the group as a whole (as in an intrapersonal prisoner's dilemma). But given the way these three effects (and probably others) can support each other in the contexts we do investigate, we think it is a mistake to try to explain all diachronic unity in one way.

11.3   Unity in the Absence of Binding Intentions

Aidan is about to have an afternoon off at home. He would like to finally reorganize his closet after some recent shopping trips, which would take all afternoon to complete. But there is also a football game playing all afternoon that he knows some of his friends will want to talk about tomorrow. He is not terribly enthusiastic about football, but he does sometimes enjoy watching a game. He knows that there will be some moments of the afternoon when he is more interested in watching football and some moments when he is more interested in organizing his closet. Every minute he spends doing one or the other of these activities will be a minute well spent doing something he cares about to some degree, but he may change his mind about which is more valuable. However, at any moment he will think it at least as good to have spent all afternoon doing just one of the two (rather than splitting his time) – i.e., even at moments when he would prefer an afternoon watching football, he will think that a fully organized closet is at least as good as a half-organized closet and half a football game, and even at moments when he would prefer an entire afternoon of cleaning, he will think that watching the whole game is at least as good as watching half and organizing half the closet. But he is likely to change his mind at various points about whether watching the full game (or nearly the full game) is better than organizing the full closet (or nearly the full closet).

What should Aidan do? Is he diachronically unified? Would he benefit from forming a binding future-directed intention? We think that the following is a useful model for considering Aidan's case.

Treat the succession of time slices of Aidan as individual agents, each confronted with the decision to spend the next minute engaged in action

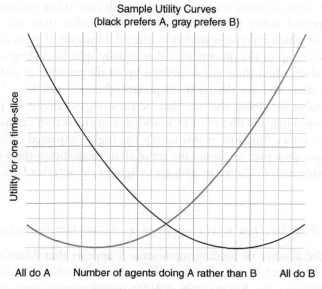

Figure 11.1 Sample utility curves

type *A* (organizing) or action type *B* (watching football).[8] Because each time slice's preferences depend on what the other time slices do, each "agent" has a utility function that is sensitive to the total number of agents that choose *A* and the total number that choose *B*. Also, because each time slice is equally ignorant about what the others prefer (except insofar as the preferences of earlier time slices are revealed by their behavior), the individual "agents" should be modeled as symmetrically uncertain about what the other time slices prefer (Figure 11.1).

When Aidan's time slices are modeled in this way, we prove the following results (see Appendix 11A). For each concave-up utility function (like those specified earlier),[9] there is an *i* such that if the first *i* time slices have

---

[8] Because we use committees of individual agents (or time slices) to model individual diachronic agency, it is worth considering whether our model can be extended to the case of *collective* agency, where the individual agents are individual humans rather than individual time slices. (See Weirich (2009) for an interesting discussion of the relationship between diachronic agency and collective agency.) We conjecture that our discussion of when and to what extent time slices qualify as unified diachronic agents is sufficiently general to apply to the questions of when and to what extent collections of individuals qualify as unified collective agents, but space limitations require us to leave this for later.

[9] A function is said to be concave up if for any two inputs, the average of the corresponding outputs is greater than the output for the average of those two inputs. For a utility function with this feature, there is some value to being more extreme rather than middling. We describe the functions in detail in the appendices.

all already done the same act type, an agent with that utility function has higher expected utility for continuing the streak rather than switching, regardless of which act type he prefers in general. However, if the actions of the time slices up to the present are either equally split (including for the first time slice, when there are no actions up to the present) or only one more has done the act the agent disprefers, then he will have higher expected value for doing the one that he prefers generally, even if that involves switching. Finally, prior to knowledge of the action of any time slice, every utility function in this family has higher expected utility for allowing one time slice to choose the act of all of them dictatorially rather than allowing each to choose its own act in light of the acts that have come before.[10]

II.3.1   What Do These Results Suggest about Aidan's Diachronic Unity?

Even in the absence of binding intentions, it appears that Aidan's time slices can be bound by his prior time slices, given the particular shape of their utility functions. For example, if Aidan spends enough time organizing the closet (or watching the game), then the rest of his time slices will be in agreement with his former time slices that it is rational to continue organizing the closet (or watching the game) despite the fact that they may well have a preference for the alternative activity, simply because they are choosing in the environment of their former time slices. Thus Aidan's earlier actions can force his later time slices to be in agreement about the rational course of action even in the absence of binding future-directed intentions. Put differently, once Aidan has established himself sufficiently on one action, he will achieve a simulacrum of self-governance just from his momentary preferences.

Does this mean that binding intentions would have no use to Aidan? No. In the beginning of the afternoon, if Aidan oscillates between preferring football and organizing (as he might, depending on the distribution of the two utility functions across the early time slices), then he will switch back and forth a few times, and if he's unlucky enough, he might spend all afternoon switching. The *preference for unanimity* result (described in Appendix 11A) implies that every time slice of Aidan would expect to be happier were Aidan able to form a binding intention at the outset to spend all afternoon doing a single one of the activities. So it seems that it would be

---

[10] One might assimilate these binding intentions to McClennen's (1990) conception of resolute choice.

beneficial to Aidan to be able to form a binding intention, even though it is possible for Aidan to be unified in the absence of such intentions.

Of course, our momentary utilities are not always (or even usually) concave up like Aidan's, so we can ask whether there are cases where agents can be unified in other ways.[11] The next example teaches us that there are, in fact, other ways that diachronic unity can be achieved solely through one's momentary preferences.

11.4 The Cost of Switching

Brooklyn is playing at the playground. On one end of the playground is a sandbox. On the other end of the playground is a swing set. It takes Brooklyn a minute to cross the playground to get from one to the other. She has 60 minutes to play at the playground. At some points in time, she values time spent on the swing set more than time spent in the sandbox. At other moments in time, this is the reverse. However, at every moment in time, she prefers time spent on the swing set or in the sandbox to time spent crossing the playground. Unlike Aidan, Brooklyn doesn't care about completeness of either activity – no matter what moment's preferences we consider, each minute spent doing one activity adds just as much utility as each other minute spent doing that activity.

Again, we consider a formal model where each time slice is considered to be a separate agent, each has her own utility function, and each is symmetrically uncertain about the utilities of each other time slice. Again, we show similar results. For each utility function of this form, there is some i such that if there are fewer than i time slices left in the sequence, she will keep doing whatever action her predecessor did rather than switching. Earlier in the process, the individual preferences will matter, but only if they are sufficiently strong compared with the cost of switching and the relative preferences of the other time slices (see Appendix 11B). Finally, prior to knowledge of the action of any time slice, every utility function in this family has higher expected utility for allowing one time slice to choose the act of all of them dictatorially rather than allowing each to choose its own act in light of the acts that have come before.

---

[11] For example, we can use concave-down utilities to model an agent who achieves diachronic self-governance by creating an environment in which her later time slices will rationally prefer to adhere to what Bratman has called a "sampling plan." In these cases, every time slice prefers heterogeneity across time (rather than homogeneity), and the later time slices will sometimes opt to do whatever the earlier time slices haven't done, even when doing so involves settling on doing something that they otherwise would not prefer.

The conditions under which one of Brooklyn's time slices will act as if bound by her predecessor are different from the conditions under which one of Aidan's time slices will act as if bound by his predecessor. But both of them will sometimes act as if bound, and every time slice of each of them would antecedently prefer that the first time slice had the capability to form a binding intention. Furthermore, increasing the cost to switching in Brooklyn's case will result in more time slices acting as if bound, and we conjecture that the same is true for increasing the concavity of Aidan's preferences.

Brooklyn, like Aidan, can be unified to some extent in the absence of binding intentions because the rational choice of a given time slice will sometimes be forced to agree with prior time slices.[12] That is, even in the absence of binding intentions, Brooklyn's choice to play in the sandbox at t can have the effect of making it rational to play in the sandbox at $t + 1$ even if Brooklyn would prefer to swing on the swing set at $t + 1$ were she to start afresh. This may make it seem as though binding intentions have no use to Brooklyn (since her prior time slices succeed at persuading her subsequent time slices sans any costs of reconsideration), but this is not right. The *preference for unanimity* result shows that all of Brooklyn's time slices would, prior to any action, prefer to have their behavior governed by a dictator with a master plan about when Brooklyn should do what. So it seems that Brooklyn's case, like Aidan's, helps us to understand why there are particular circumstances in which it is good to be diachronically unified by binding intentions.

11.5 The Costs of Introspecting

Aidan and Brooklyn teach us that it is possible for agents' earlier time slices to govern their later time slices even in the absence of binding intentions.

---

[12] If we consider a case that has features of both Brooklyn's and Aidan's, where the utility function for each time slice is concave up in the total number of time slices that do a given act, and there is *also* a cost based on the number of switches, then we can sometimes get even greater degrees of unity. For instance, we noted that in a case where the utility function was concave up and the utility of the 50/50 split was at least as great as the utility of the opposite extreme, an agent will have perfect unity if by chance, some initial sequence of time-slices all have the same preference, and the agent will have total disunity if by chance, consecutive time-slices all have opposite preferences. If a cost to switching actions is introduced, then toward the end of the sequence, the cost of switching will outweigh the possible benefit of having more of one's preferred action, so even the maximally disunified sequence will exhibit some unity in its behavior. However, for the sequence that is already maximally unified, this cost doesn't add any more unity. A detailed analysis of when the two factors together produce more unity than either does separately would require further investigation of specific utility functions. But even with what we have now, we can see that maximal unity is possible either with a high cost to switching, or with sufficiently concave utilities and a sufficient streak of agreement in the early stages, or with some combination.

By initially settling on some particular course of action, Aidan's and Brooklyn's early time slices effectively create an environment in which it is optimal for Aidan's and Brooklyn's later time slices to settle on the same course of action, even though Aidan's and Brooklyn's later time slices would otherwise prefer to do something else.

Although these cases demonstrate that there are multiple ways to achieve diachronic unity, they do not show that an agent's earlier time slices can govern her later time slices without changing what the later time slices want to do.[13] This means that these cases do not shed light on the kind of rational self-governance that exists when Carmen is rationally required to stick with her plan to have the apple because she takes there to be costs to determining what she wants but remains neutral between the apple and the orange.

In order to model cases such as Carmen's, we must introduce the costs and benefits of introspecting. On doing so, it is clear that the rationality of *x*-ing *on introspecting* must be distinguished from the rationality of *x*-ing *absent introspection* because the rationality of these actions can come apart. For example, when Carmen believes that there are perks to thinking about what fruit she wants before her friend arrives, it seems that Carmen prefers introspecting and having an apple (or an orange) to not introspecting and having an apple and that if Carmen is rational, she will either introspect and have the apple or introspect and have the orange.

There may be multiple ways to account for the costs and benefits of introspecting. Here we propose modeling the cost of introspecting as a price that the agent pays in order to infer what she prefers in the moment or, put differently, in order to remove any ignorance about her utility function.[14] Thus we model the agent who does not go to the trouble to reopen a settled question as ignorant about what course of action she regards as optimal in the moment but as possibly paying a price to infer whether she finds sticking with her prior plan optimal (see Appendix 11C).

This cost of introspecting provably can render the agent bound by her prior plans. For example, if Carmen treats her prior plan to have the apple as evidence that she currently wants the apple, then it is rational for her to

---

[13] One may worry that we have not described cases of genuine self-*governance* because each time slice just does what it most wants to do. By our lights, this reply requires too narrow of a conception of governance. For example, an employer clearly *governs* her employees inasmuch as she gives them incentive to prefer taking some actions to others. This is basically what Aidan and Brooklyn do, since they promote an environment in which their later selves will be motivated to serve their earlier interests.

[14] In the formal model, we model Carmen as having one known utility function but as being uncertain about her own psychology.

stick with her prior plan if the costs of introspecting are sufficiently high.[15] Alternatively, if Carmen thinks (absent introspection) that she probably prefers the apple to the orange because there is usually a cost to deviating from standing plans, then even if Carmen, as a matter of fact, is neutral between the two options, it can be rational to stick with the prior plan in order to avoid the costs of introspecting. In contrast, if Carmen were to think that there was no cost (or even a benefit) to discovering her preferences, then Carmen would have no reason (or even a negative reason) to abstain from determining what she wants and would therefore be rationally required to opt for some course of action that she momentarily (actually) regards as among the optimal actions. Thus, when Carmen is actually neutral between the apple and the orange and thinks that she stands to benefit from introspecting (because it helps to alleviate the boredom), she can rationally opt for either the apple or the orange.

As we discuss in Appendix 11C, if we revise the cases of Aidan and Brooklyn to introduce costs of introspection, then Aidan and Brooklyn will be even more diachronically unified. There is also a formal similarity between the way that a switching cost and a cost to introspection induce unity, although the cost to introspection turns out to be more effective.

11.6   The Costs of Reassessing

At this juncture we have demonstrated that there are multiple ways to achieve diachronic unity in particular circumstances without any recourse to diachronic norms of coherence. But we have not modeled *every* way that agents can achieve diachronic unity. For example, the following case evades our current grasp.

At the beginning of an offensive basketball possession, Diana settles on driving left toward the basket. As she gets closer to the basket, it may be best (in expectation) for Diana to stick to her guns and not reconsider her prior intention (perhaps because slowing down to consider what seems best often leads to turnovers), but in the event that she does reconsider, it is better for Diana to go right than left (perhaps because she now judges the path to the right to be more open).

Here it seems that Diana should not reconsider her prior choice, but not because of any price associated with introspecting. Indeed, Diana may know perfectly well when driving left is optimal and when driving right is

---

[15] This is true even though Carmen would discover that she is neutral between the apple and the orange were she to introspect.

optimal; it just takes Diana time and effort (i.e., better spent driving toward the basket) to determine whether her current circumstances call for driving left or driving right. It is natural to model Diana's earlier time slice as an expert that Diana's later time slice can defer to when she does not take the time to investigate the circumstances herself. Were Diana to take the time to survey her circumstances once more, she'd discover that it's better to change courses. But because doing so takes time and effort, it can still be rational for Diana to defer to the prior time slice's assessment by continuing left.[16]

Although the costs of introspecting can be modeled as a price that one can pay to eliminate uncertainty about what one desires, this exact treatment is not available here because Diana is not uncertain about what she values. However, if we slightly modify the existing model so that Diana can pay a price to reassess the state of play (rather than deferring to her past time slice), then we can capably represent Diana's choice.[17] The unity here may not be of the same kind discussed in previous cases because it resists changing external circumstances, not internal circumstances, but it does seem that both binding intentions and models such as ours can capture the kind of self-governance at play here. Indeed, this model may be interpreted as a way to represent Bratmanian binding intentions.

11.7 Lessons Learned

Aidan, Brooklyn, and Carmen teach us that there are ways for agents to be diachronically unified (and self-governed) even in the absence of binding intentions, but they also reveal that there are contexts in which diachronic agents can be better off (in expectation) for having binding intentions.

On the one hand, this means that there really is something remarkable about having the capacity to form binding intentions (or to execute plans over time) because collections of time slices that have this capacity can outperform those that do not. On the other hand, this suggests that there are multiple ways to achieve diachronic unity even in the absence of binding intentions and that we cannot be sure that an agent has exhibited self-governance (by forming a binding intention) just from observing her

---

[16] One might argue that it is wrong to include the option of not reassessing the circumstances because Diana does not actively decide whether to reassess as she drives toward the basket. Although we agree that agents usually are not confronted with conscious choices about whether to reassess (or to introspect, for that matter), we believe that we can nevertheless use an agent's probabilities and utilities to *evaluate* whether she was rational to reassess or introspect.

[17] We discuss this in greater detail in the appendices.

diachronic unity because her unity may have resulted from her momentary preferences being aligned in the right way.[18]

Either way, whether an agent should follow through with her prior plans seems to depend on expected utility considerations in a great many circumstances (perhaps excepting cases like Diana's). So whether an agent should follow through with her plans depends on how costly (or beneficial) it is to do otherwise and on how much she stands to gain from taking the steps necessary to steer herself in a new direction.

APPENDIX 11A

Aidan: Concave-Up Utilities

Aidan has an afternoon to spend either organizing his closet or watching a football game. He might change his mind about which activity he prefers over the course of the afternoon, but our formal model shows that he may nevertheless make a decision and stick with it. We consider him as a sequence of time slices and treat each time slice as an independent agent with preferences regarding the collective behavior of all time slices, each given by its own utility function. Importantly, although the time slices may disagree about which activity is better, they all agree that completion is preferred over splitting time.

A.1   Formal Assumptions

There is a fixed finite number n of agents who are to make decisions, and each agent is certain about how many agents there are.

Every agent has two choices, A and B. Each agent has a utility function that depends only on the number a of agents that choose A and the number $b = n - a$ of agents that choose B. For each agent, the utility function is concave up – for any i and j, the utility when $a = (i + j)/2$ is less than the average of the utility when $a = i$ and the utility when $a = j$. For each agent, the utility of an even split between A and B is at least as high as the utility of one type of unanimity and is less than the other type of unanimity. One such utility function is given by $a^2 + 3b^2$, but many others are possible.

Every agent is certain of the preceding facts for the utility function of each other agent, but none of them knows each other's utility functions.

---

[18] This suggests that Martin might have been right to be impressed with Zeb's seeming capacity to form binding intentions but also might have been right not to be certain that Zeb really had this capacity.

At the time of action, each agent is certain of the behavior of each earlier agent but has no information about the behavior of any later one. Prior to observing any behavior, each agent has independent, identically distributed credences about the utility function of each other agent, and these credences are symmetrical with respect to interchanges of A and B.

This could be because there are just two utility functions that are mirror images of each other, and the agent has an independent 0.5 credence that each other agent is of one of these two types, or because there are many utility functions with greater or lesser degrees of favoring of one unanimous type over the other, or greater or lesser degrees of concavity.

A.2 Results

As a result of concavity, and because the 50/50 split is at least as good as unanimity on the disfavored action, each agent also prefers a 50/50 split over any nonunanimous majority for the disfavored action. Furthermore, by concavity, any nonextreme symmetrical probability distribution over the behavior of the group has a higher expected utility than a guaranteed 50/50 split and a lower expected utility than a 0.5 credence in each extreme.

> Lemma 1: If, conditional on a particular initial sequence S of choices, the expected utility of doing act A is at least as high as that of doing act B, then conditional on an initial sequence SA, the expected utility of doing act A is still at least as high as that of doing act B. That is, regardless of whether the agent prefers A to B generally, if a given initial sequence makes act A preferable, then an additional A by one extra time slice beforehand won't change this fact.

Proof: We will prove the contrapositive – if for a given utility function, conditional on initial sequence SA, the expected value of doing B is greater than that of doing A, then conditional on initial sequence S, the expected value of doing B is greater than that of doing A.

As part of the proof, we will use this corollary: for any probability distribution over utility functions, the probability of an agent preferring B conditional on initial sequence S is at least as high as the probability of the agent preferring B conditional on initial sequence SA.

A.3   Proof by Induction on the Number of Remaining Agents after the Current One

Base case: The length of S is $n - 2$, so the act after SA is the last one. Assume that the utility function is such that the sequence SAB has at least as high utility as SAA. Each utility function under consideration is a function of b, the total number of B's in the sequence. Because the utility function is concave up, there is a number i such that the utility of a total sequence is monotonically increasing in b when $b > i$ and monotonically decreasing in b when $b < i$. Since the utility of SAB is greater than that of SAA, the number of B's in S must be at least i. Thus $U(SAA) \leq U(SAB) = U(SBA) < U(SBB)$. The expected utility of SA is a weighted average of $U(SAA)$ and $U(SAB)$, and the expected utility of SB is a weighted average of $U(SBA)$ and $U(SBB)$, where the weights are the probabilities of the last agent having a utility function that leads it to do A or B. Because $U(SBA)$ and $U(SBB)$ are at least as high as $U(SAA)$ and $U(SAB)$ and $U(SBB)$ is strictly greater than $U(SAA)$, this means that given initial sequence S, the expected value of doing B is strictly greater than that of doing A.

For the induction step, assume that we have shown for all utility functions in the relevant family that if $EU(S'AB) > EU(S'AA)$ when there are j unknown agents after the ones mentioned here, then $EU(S'B) > EU(S'A)$. Now we must show that if $EU(SAB) > EU(SAA)$ when there are $j + 1$ unknown agents after the ones mentioned, then $EU(SB) > EU(SA)$.

Note that $EU(SB)$ is a weighted average $pEU(SBA) + (1 - p)EU(SBB)$ and that $EU(SA)$ is a weighted average $qEU(SAA) + (1 - q)EU(SAB)$, where p and q are the probabilities of an unknown agent preferring act B conditional on initial sequences SB and SA, respectively. Note that $EU(SBA) = EU(SAB)$, and by the assumption of the theorem, we have $EU(SAB) > EU(SAA)$. Thus, if $EU(SBB) \geq EU(SBA)$, then we are done.

> Lemma 2: An agent's expected utility given an initial sequence S of choices, and given that all later agents have her same utility function, is at least as great as her expected utility given an initial sequence of choices, with no information about the utility functions of later agents.

Proof: The expected utility of a given sequence of choices is a mixture of the expected utility of the next agent doing A and that of the next agent doing B. Conditioning on the agent at that time having the same utility function as the agent under consideration removes all weight from whichever of these two expected utilities is lower for the agent under consideration. Thus, conditioning on every agent having the same utility function as the agent under consideration just increases the expected utility at each step.

> Proposition: Conditional on an initial sequence of choices containing i choices of the agent's favored act and $i + 1$ choices of the agent's disfavored act, the agent weakly prefers to do her favored act, with equality only if $i = 0$ and she is the last agent to act.

Proof: By Lemma 1, if a single-act majority for the disfavored act were sufficient to prefer doing the disfavored act, then so would any greater majority. Thus, if all future agents were of the same type as this agent, the actual outcome would be unanimity if $i = 0$ and a split majority for the disfavored act if $i > 0$. Thus, by Lemma 2, the actual expected utility of doing the disfavored act with no information about the type of later agents must be no greater than this, which is at most equal to the utility of a 50/50 split if $i = 0$, and strictly less if $i > 0$.

However, if the agent performs her preferred act, then the information available for the next agent is symmetrical. Thus, with no information about future agent types, her credences over complete outcomes must be symmetrical and nonextreme if she is not the last agent to act. Thus the expected utility conditional on her performing her favored act is strictly greater than that of a 50/50 split if she is not the last agent to act, and the expected utility of her performing her disfavored act is strictly less than that of a 50/50 split if $n > 0$. Thus, if either $n > 0$ or she is not the last agent to act, she prefers in this circumstance to do her preferred act.

If $n = 0$ and she is the last agent to act, then she might be indifferent between the two acts because one produces total unanimity and the other produces a 50/50 split.

Combining this proposition with Lemma 1, we see that conditional on a prior 50/50 split, an agent prefers to perform her favored act. Thus we can see that if there are more than two agents in the sequence and the first two favor different actions (no matter how much or how little they favor the different extremes), the group will not perform a unanimous action.

A.4 Preference for Unanimity Result

Thus, if there are more than two agents in the sequence, then the prior credence for each agent will be a nonextreme symmetrical distribution over action sequences, which has strictly lower expected utility than a 0.5 credence in each unanimous action sequence. Because this 0.5 credence in each unanimous action sequence is the expected result of giving the first agent in the sequence the ability to bind all future agents to act the same way as the first, each agent will in the abstract prefer the ability for the first agent to make such binding decisions, even though she would prefer to act differently from the first if she happened to go second after an agent of the opposite type.

APPENDIX 11B

Brooklyn: Linear Utility with a Switching Cost

Brooklyn has an hour to spend at the playground, splitting her time between playing in the sandbox, playing on the swing set, and running between the two. She might change her mind about which activity she prefers at any moment, but our model shows that she will sometimes stick with the dispreferred activity and may even stick with it for the entire hour despite changing her mind. Again, the formal model considers each time slice as her own agent with her own utility function and uncertainty about the utility functions of other time slices.

B.1   Assumptions

There is a fixed finite number n of agents who are to make decisions, and each agent is certain about the value of n.

There are two types of actions – A and B. For each agent, the utility function is determined by three positive constants, x, y, and s, and takes the form $ax + by - ds$, where a is the number of agents that chose A, b is the number of agents that chose B, and d is the number of agents that chose a different act from her predecessor. For each agent, the switching cost s is greater than the difference of x and y. (The assumption that the total utility of a time slice is given by her own values for the experience of each other time slice's behavior is unrealistic if these utilities are taken to be hedonic utilities – in this case it would be more plausible for the total utility of a time slice to depend on how many time slices do their *own* preferred behaviors rather than hers. But this assumption may be more plausible if the value of the play involves spinning out an elaborate fiction based on each activity, which depends on the behavior of other time slices involved in the same activity.)

Each agent is certain of these facts for all other agents, but none of them knows the specific constants of any other agent. Each has i.i.d. credences about the utility function of each other agent, and these credences are symmetrical with respect to interchanges of A and B.

Some agents will have preferences and positions in the sequence such that they will do the act that has higher utility for them regardless of what their predecessor did, whereas others will have preferences and positions in the sequence such that they will do what their predecessor did, regardless of which act has higher utility for them. If there is a consecutive sequence of agents of the latter type, we say that they are all *controlled* by the last agent to choose on the basis of her own preferences before that sequence.

Consider an agent about to decide whether to be controlled by her predecessor or to choose on the basis of her own preference. Let c_i be the number of time slices (counting herself) that are expected to be controlled by the choice of the ith-to-last agent if she chooses for herself. $c_i|b - a|$ is then the expected advantage of choosing based on her own preference rather than choosing the opposite of her own preference, while s is the expected advantage of choosing the act chosen by her predecessor rather than the opposite. Thus the ith-to-last agent will be controlled by her predecessor iff $c_i|b - a| < s$.

Let $p(c)$ be the probability that a randomly selected agent has $c|b - a| < s$. Then $c_{i+1} = 1 + c_i p(c_i)$ because the agent in position $i + 1$ from the end would expect to control herself and have probability $p(c_i)$ of controlling the c_i that would be controlled by her successor. Since $p(c)$ is the probability that a randomly selected agent has utility function with $s/|b - a| > c$, we see that $p(0) = 1$ (because s is always positive), and p monotonically decreases to 0 as c goes to infinity.

Let $t = s/|b - a|$. This is the threshold for c at which a given agent would be willing to choose according to her own preference rather than according to the act the previous agent chose. If everyone has the same value of t, then the tth-to-last agent is the last one that is willing to choose according to her preference, and all later agents will be controlled by her. Since the tth-to-last agent controls herself, the previous agent will act as if she is the last, and by parallel reasoning, every block of t agents, counting from the end, will be controlled by its first member. Thus, in this situation, there is a very real sense in which we could treat the sequences of t time slices as the units rather than the time slices themselves.

If agents don't have the same value of t but instead it is distributed continuously, then $p(c)$ is a continuous function, so $1/[1 - p(C)]$ is a continuous, monotonically decreasing function going to infinity as

C goes to o and going to o as C goes to infinity. Thus there is one unique value for which $C = 1/[1 - p(C)]$ or, equivalently, $1 + Cp(C) = C$. As i gets large, c_i will converge to this value. Thus, as long as it is far from the end of the sequence, everyone whose t is in the most extreme $1/C$ of the population will choose according to her own preferences, whereas everyone whose t is less extreme will be controlled by her predecessor. However, in the last few segments of the sequence, even agents whose preferences are this extreme will be controlled by their predecessor if there are not enough successors left to make up for her t.

Note that as s gets large while the distribution of $|b - a|$ stays fixed, C gets larger, and a larger fraction of agents will be controlled by their predecessor. Thus higher switching costs translate to greater behavioral unity.

B.2   Preference for Unanimity Result

The expected value of being bound by the first agent is equal to the average of the two unanimous sequences. The expected value of not being bound is the average of some symmetrical distribution over the sequences. The average of the value of any sequence and its dual is equal to the average of the two unanimous sequences minus s times the number of switches that each sequence has. Thus, again, every agent will prefer that the collective be bound by the first agent, even though some agents might prefer to switch.

Note that if there is an upper bound to the possible value of $|b - a|$, then if s gets high enough, there will be no switches, and this can mimic a binding intention. So increases in the cost of switching with no compensating increase in the payoff for any act can paradoxically increase the expected utility of time slices. However, this increase isn't always monotonic.

APPENDIX 11C

Carmen: Costs to Introspection

Carmen is choosing whether to take an apple or an orange and is uncertain which she prefers. She can spend some effort introspecting to get certainty about her preferences or just act according to her uncertain preferences.

For the formal version of this, we will start by considering a one-shot version of the decision rather than a sequence of time slices, as in the examples of Aidan and Brooklyn. We will provide some discussion of how this effect of a cost to introspection changes things if added to sequential choice cases, but we won't provide a full analysis. This is because the analysis is fairly uninteresting if every time slice has the same symmetrical uncertainty, but if they don't, then the analysis depends on uncertainties not just about each other's utility functions but also about each others uncertainties, and the complexity quickly gets out of hand.

For a formal analysis, we can treat Carmen as having a choice between two actions, A and B, and uncertainty about her own utility for each action. Under function U_1, A has greater utility than B, but under function U_2, B has greater utility than A, and Carmen is uncertain which utility function is actually hers. We can formally model this as though Carmen has one *known* utility function and an *unknown* bit of information about her own psychological state so that $U_1(A)$ is treated as $U(x_1, A)$, $U_2(A)$ is treated as $U(x_2, A)$, and so on. If p is the probability that she has function U_1, then we can see that the choice to do act A has expected utility $pU(x_1, A) + (1 - p)U(x_2, A)$, and the choice to do act B has expected utility $pU(x_1, B) + (1 - p)U(x_2, B)$.

If Carmen has the option to introspect, for introspection cost i, then she can use this introspection to do whichever act in fact has higher utility for her. The expected value of this plan is $pU(x_1, A) + (1 - p)U(x_2, B) - i$. She will rationally choose to pay this cost iff $i < p[U(x_1, A) - U(x_1, B)]$ and $i < (1 - p)[U(x_2, B) - U(x_2, A)]$. If i is higher than either threshold, though, she will just choose to do whichever act appears to have higher expected utility in light of her uncertainty. If U_1 and U_2 are symmetrical with respect to A and B, then this means that she will rationally pay the cost of introspection iff the probability of the less likely utility function multiplied by the difference in utility of the two outcomes is greater than the introspection cost, and otherwise she will rationally just do whichever act has the higher probability of being better. (When Carmen is bored, the introspection "cost" is negative because she derives entertainment from introspection, so she will definitely pay the cost.) Thus, if an earlier plan is taken as evidence that one's utility function is more likely to still favor that act than the other, Carmen will rationally stick with this plan unless the possible difference in values between the two outcomes is especially large.

If we compare this to a one-shot version of Brooklyn's case, where she can pay a switching cost s to do B or do A for free, we see that Brooklyn will switch iff $s < [U(x_2, B) - U(x_2, A)]$, whereas Carmen will switch iff $i < (1 - p)[U(x_2, B) - U(x_2, A)]$. The difference between the two is a multiplicative factor of the probability of the less likely preference, so a relatively low introspection cost may well be more effective than a somewhat higher switching cost at producing behavior such as what we saw in the analysis of Brooklyn's case earlier (a fuller analysis of an iterated version of this case would require some theory of how the probabilities of the two utility functions evolve over time and an analysis of how many future time slices the combination of switching cost and introspection cost allows one time slice to control).

Note that a formally similar analysis is available for Diana's case if we reinterpret x_1 and x_2 not as internal psychological states that determine the value of the two acts but rather as external states of the game that determine the value of driving left or right and interpret i not as the cost of intro-specting which psychological state one is in but instead as the cost of analyzing the state of play to determine which direction would be better to drive in. If an earlier decision to drive left is evidence that the game is probably still in a state where driving left is better, then Diana will continue to drive left unless the uncertainty about which direction is better is high enough and the value of the better direction great enough to overcome the cost of analyzing the game.

Combining this notion of an introspection cost with Aidan's feature of concave-up utilities is even more complex, but it can be worked out in a simple case, to see that even a relatively small introspection cost can change things. Consider a case with just three time slices, each having a utility function of either $a^2 + 3b^2$ or $3a^2 + b^2$. If the first two time slices have done the same act and the third is debating which to do, then his choice is either between 9 or 7 units of utility or between 27 or 13, depending on which utility function he has. In any case, he would prefer doing the same as the previous two. If the first two time slices have done different acts, then the third is choosing between 13 and 7 units of utility. If the introspection cost i is less than 3 units of utility, he will pay it, but if $i > 3$, then he will choose arbitrarily, if he thinks either preference is equally likely. If we consider the choice facing the second agent, he can either do the same act as the first agent or a different one. If he does the same one, then the third agent will go along with the choice, so the overall payoff will be either 27 or 9. If he does the other one, then the third agent will either choose arbitrarily (if $i > 3$) or introspect and then choose in line with that (which is just as likely to agree with his current preferences or go against them). Doing the opposite action is only going to have higher expected utility than going along with the first agent if $i < \frac{1}{2}$ and the second agent introspects and recognizes that his preference goes against the act that the first agent did.

In a sense, the fact that a cost to introspection leads to unified behavior should be the least surprising result of these formal models because the agent who hasn't yet introspected is making a decision on behalf of a mixture of her own utility function and the utility functions of other time slices rather than merely on her own behalf.

References

Bratman, M. 1987. *Intentions, Plans, and Practical Reasons*. Cambridge, MA: Harvard University Press.

Bratman, M. 2012. Time, rationality, and self-governance. *Philosophical Issues* 22:73–88.

Chang, R. 2002. The possibility of parity. *Ethics* 112:659–88.

McClennen, E. 1990. *Rationality and Dynamic Choice*. Cambridge: Cambridge University Press.

Weirich, P. 2009. *Collective Rationality*. Oxford: Oxford University Press.

Index

264